The Scientific Revolution: A Very Short Introduction

VERY SHORT INTRODUCTIONS are for anyone wanting a stimulating and accessible way in to a new subject. They are written by experts and have been translated into more than 40 different languages. The series began in 1995 and now covers a wide variety of topics in every discipline. The VSI library contains nearly 400 volumes—a Very Short Introduction to everything from Indian philosophy to psychology and American history—and continues to grow in every subject area.

Very Short Introductions available now:

CAUSATION Stephen Mumford and Rani Lill Anjum
THE CELL Terence Allen and Graham Cowling
THE CELTS Barry Cunliffe
CHAOS Leonard Smith
CHILDREN'S LITERATURE Kimberley Reynolds
CHINESE LITERATURE Sabina Knight
CHOICE THEORY Michael Allingham
CHRISTIAN ART Beth Williamson
CHRISTIAN ETHICS D. Stephen Long
CHRISTIANITY Linda Woodhead
CITIZENSHIP Richard Bellamy
CIVIL ENGINEERING David Muir Wood
CLASSICAL LITERATURE William Allan
CLASSICAL MYTHOLOGY Helen Morales
CLASSICS Mary Beard and John Henderson
CLAUSEWITZ Michael Howard
CLIMATE Mark Maslin
THE COLD WAR Robert McMahon
COLONIAL AMERICA Alan Taylor
COLONIAL LATIN AMERICAN LITERATURE Rolena Adorno
COMEDY Matthew Bevis
COMMUNISM Leslie Holmes
COMPLEXITY John H. Holland
THE COMPUTER Darrel Ince
THE CONQUISTADORS Matthew Restall and Felipe Fernández-Armesto
CONSCIENCE Paul Strohm
CONSCIOUSNESS Susan Blackmore
CONTEMPORARY ART Julian Stallabrass
CONTEMPORARY FICTION Robert Eaglestone
CONTINENTAL PHILOSOPHY Simon Critchley
CORAL REEFS Charles Sheppard
COSMOLOGY Peter Coles
CRITICAL THEORY Stephen Eric Bronner
THE CRUSADES Christopher Tyerman
CRYPTOGRAPHY Fred Piper and Sean Murphy
THE CULTURAL REVOLUTION Richard Curt Kraus
DADA AND SURREALISM David Hopkins
DARWIN Jonathan Howard
THE DEAD SEA SCROLLS Timothy Lim
DEMOCRACY Bernard Crick
DERRIDA Simon Glendinning
DESCARTES Tom Sorell
DESERTS Nick Middleton
DESIGN John Heskett
DEVELOPMENTAL BIOLOGY Lewis Wolpert
THE DEVIL Darren Oldridge
DIASPORA Kevin Kenny
DICTIONARIES Lynda Mugglestone
DINOSAURS David Norman
DIPLOMACY Joseph M. Siracusa
DOCUMENTARY FILM Patricia Aufderheide
DREAMING J. Allan Hobson
DRUGS Leslie Iversen
DRUIDS Barry Cunliffe
EARLY MUSIC Thomas Forrest Kelly
THE EARTH Martin Redfern
ECONOMICS Partha Dasgupta
EDUCATION Gary Thomas
EGYPTIAN MYTH Geraldine Pinch
EIGHTEENTH-CENTURY BRITAIN Paul Langford
THE ELEMENTS Philip Ball
EMOTION Dylan Evans
EMPIRE Stephen Howe
ENGELS Terrell Carver
ENGINEERING David Blockley
ENGLISH LITERATURE Jonathan Bate
ENVIRONMENTAL ECONOMICS Stephen Smith
EPIDEMIOLOGY Rodolfo Saracci
ETHICS Simon Blackburn
THE EUROPEAN UNION John Pinder and Simon Usherwood
EVOLUTION Brian and Deborah Charlesworth
EXISTENTIALISM Thomas Flynn
THE EYE Michael Land
FAMILY LAW Jonathan Herring
FASCISM Kevin Passmore
FASHION Rebecca Arnold
FEMINISM Margaret Walters
FILM Michael Wood
FILM MUSIC Kathryn Kalinak
THE FIRST WORLD WAR Michael Howard
FOLK MUSIC Mark Slobin
FOOD John Krebs
FORENSIC PSYCHOLOGY David Canter
FORENSIC SCIENCE Jim Fraser
FOSSILS Keith Thomson

FOUCAULT Gary Gutting
FRACTALS Kenneth Falconer
FREE SPEECH Nigel Warburton
FREE WILL Thomas Pink
FRENCH LITERATURE John D. Lyons
THE FRENCH REVOLUTION William Doyle
FREUD Anthony Storr
FUNDAMENTALISM Malise Ruthven
GALAXIES John Gribbin
GALILEO Stillman Drake
GAME THEORY Ken Binmore
GANDHI Bhikhu Parekh
GENIUS Andrew Robinson
GEOGRAPHY John Matthews and David Herbert
GEOPOLITICS Klaus Dodds
GERMAN LITERATURE Nicholas Boyle
GERMAN PHILOSOPHY Andrew Bowie
GLOBAL CATASTROPHES Bill McGuire
GLOBAL ECONOMIC HISTORY Robert C. Allen
GLOBAL WARMING Mark Maslin
GLOBALIZATION Manfred Steger
THE GOTHIC Nick Groom
GOVERNANCE Mark Bevir
THE GREAT DEPRESSION AND THE NEW DEAL Eric Rauchway
HABERMAS James Gordon Finlayson
HAPPINESS Daniel M. Haybron
HEGEL Peter Singer
HEIDEGGER Michael Inwood
HERODOTUS Jennifer T. Roberts
HIEROGLYPHS Penelope Wilson
HINDUISM Kim Knott
HISTORY John H. Arnold
THE HISTORY OF ASTRONOMY Michael Hoskin
THE HISTORY OF LIFE Michael Benton
THE HISTORY OF MATHEMATICS Jacqueline Stedall
THE HISTORY OF MEDICINE William Bynum
THE HISTORY OF TIME Leofranc Holford-Strevens
HIV/AIDS Alan Whiteside
HOBBES Richard Tuck
HORMONES Martin Luck
HUMAN EVOLUTION Bernard Wood
HUMAN RIGHTS Andrew Clapham
HUMANISM Stephen Law
HUME A. J. Ayer
HUMOUR Noël Carroll
THE ICE AGE Jamie Woodward
IDEOLOGY Michael Freeden
INDIAN PHILOSOPHY Sue Hamilton
INFORMATION Luciano Floridi
INNOVATION Mark Dodgson and David Gann
INTELLIGENCE Ian J. Deary
INTERNATIONAL MIGRATION Khalid Koser
INTERNATIONAL RELATIONS Paul Wilkinson
INTERNATIONAL SECURITY Christopher S. Browning
ISLAM Malise Ruthven
ISLAMIC HISTORY Adam Silverstein
ITALIAN LITERATURE Peter Hainsworth and David Robey
JESUS Richard Bauckham
JOURNALISM Ian Hargreaves
JUDAISM Norman Solomon
JUNG Anthony Stevens
KABBALAH Joseph Dan
KAFKA Ritchie Robertson
KANT Roger Scruton
KEYNES Robert Skidelsky
KIERKEGAARD Patrick Gardiner
THE KORAN Michael Cook
LANDSCAPE ARCHITECTURE Ian H. Thompson
LANDSCAPES AND GEOMORPHOLOGY Andrew Goudie and Heather Viles
LANGUAGES Stephen R. Anderson
LATE ANTIQUITY Gillian Clark
LAW Raymond Wacks
THE LAWS OF THERMODYNAMICS Peter Atkins
LEADERSHIP Keith Grint
LINCOLN Allen C. Guelzo
LINGUISTICS Peter Matthews
LITERARY THEORY Jonathan Culler
LOCKE John Dunn
LOGIC Graham Priest
MACHIAVELLI Quentin Skinner
MADNESS Andrew Scull
MAGIC Owen Davies
MAGNA CARTA Nicholas Vincent

MAGNETISM Stephen Blundell
MALTHUS Donald Winch
MANAGEMENT John Hendry
MAO Delia Davin
MARINE BIOLOGY Philip V. Mladenov
THE MARQUIS DE SADE John Phillips
MARTIN LUTHER Scott H. Hendrix
MARTYRDOM Jolyon Mitchell
MARX Peter Singer
MATHEMATICS Timothy Gowers
THE MEANING OF LIFE Terry Eagleton
MEDICAL ETHICS Tony Hope
MEDICAL LAW Charles Foster
MEDIEVAL BRITAIN John Gillingham and Ralph A. Griffiths
MEMORY Jonathan K. Foster
METAPHYSICS Stephen Mumford
MICHAEL FARADAY Frank A.J.L. James
MICROECONOMICS Avinash Dixit
MODERN ART David Cottington
MODERN CHINA Rana Mitter
MODERN FRANCE Vanessa R. Schwartz
MODERN IRELAND Senia Pašeta
MODERN JAPAN Christopher Goto-Jones
MODERN LATIN AMERICAN LITERATURE Roberto González Echevarría
MODERN WAR Richard English
MODERNISM Christopher Butler
MOLECULES Philip Ball
THE MONGOLS Morris Rossabi
MORMONISM Richard Lyman Bushman
MUHAMMAD Jonathan A.C. Brown
MULTICULTURALISM Ali Rattansi
MUSIC Nicholas Cook
MYTH Robert A. Segal
THE NAPOLEONIC WARS Mike Rapport
NATIONALISM Steven Grosby
NELSON MANDELA Elleke Boehmer
NEOLIBERALISM Manfred Steger and Ravi Roy
NETWORKS Guido Caldarelli and Michele Catanzaro
THE NEW TESTAMENT Luke Timothy Johnson
THE NEW TESTAMENT AS LITERATURE Kyle Keefer
NEWTON Robert Iliffe
NIETZSCHE Michael Tanner
NINETEENTH-CENTURY BRITAIN Christopher Harvie and H. C. G. Matthew
THE NORMAN CONQUEST George Garnett
NORTH AMERICAN INDIANS Theda Perdue and Michael D. Green
NORTHERN IRELAND Marc Mulholland
NOTHING Frank Close
NUCLEAR POWER Maxwell Irvine
NUCLEAR WEAPONS Joseph M. Siracusa
NUMBERS Peter M. Higgins
NUTRITION David A. Bender
OBJECTIVITY Stephen Gaukroger
THE OLD TESTAMENT Michael D. Coogan
THE ORCHESTRA D. Kern Holoman
ORGANIZATIONS Mary Jo Hatch
PAGANISM Owen Davies
THE PALESTINIAN-ISRAELI CONFLICT Martin Bunton
PARTICLE PHYSICS Frank Close
PAUL E. P. Sanders
PENTECOSTALISM William K. Kay
THE PERIODIC TABLE Eric R. Scerri
PHILOSOPHY Edward Craig
PHILOSOPHY OF LAW Raymond Wacks
PHILOSOPHY OF SCIENCE Samir Okasha
PHOTOGRAPHY Steve Edwards
PLAGUE Paul Slack
PLANETS David A. Rothery
PLANTS Timothy Walker
PLATO Julia Annas
POLITICAL PHILOSOPHY David Miller
POLITICS Kenneth Minogue
POSTCOLONIALISM Robert Young
POSTMODERNISM Christopher Butler
POSTSTRUCTURALISM Catherine Belsey
PREHISTORY Chris Gosden
PRESOCRATIC PHILOSOPHY Catherine Osborne
PRIVACY Raymond Wacks
PROBABILITY John Haigh
PROGRESSIVISM Walter Nugent
PROTESTANTISM Mark A. Noll

PSYCHIATRY Tom Burns
PSYCHOLOGY Gillian Butler and Freda McManus
PURITANISM Francis J. Bremer
THE QUAKERS Pink Dandelion
QUANTUM THEORY John Polkinghorne
RACISM Ali Rattansi
RADIOACTIVITY Claudio Tuniz
RASTAFARI Ennis B. Edmonds
THE REAGAN REVOLUTION Gil Troy
REALITY Jan Westerhoff
THE REFORMATION Peter Marshall
RELATIVITY Russell Stannard
RELIGION IN AMERICA Timothy Beal
THE RENAISSANCE Jerry Brotton
RENAISSANCE ART Geraldine A. Johnson
REVOLUTIONS Jack A. Goldstone
RHETORIC Richard Toye
RISK Baruch Fischhoff and John Kadvany
RIVERS Nick Middleton
ROBOTICS Alan Winfield
ROMAN BRITAIN Peter Salway
THE ROMAN EMPIRE Christopher Kelly
THE ROMAN REPUBLIC David M. Gwynn
ROMANTICISM Michael Ferber
ROUSSEAU Robert Wokler
RUSSELL A. C. Grayling
RUSSIAN HISTORY Geoffrey Hosking
RUSSIAN LITERATURE Catriona Kelly
THE RUSSIAN REVOLUTION S. A. Smith
SCHIZOPHRENIA Chris Frith and Eve Johnstone
SCHOPENHAUER Christopher Janaway
SCIENCE AND RELIGION Thomas Dixon
SCIENCE FICTION David Seed
THE SCIENTIFIC REVOLUTION Lawrence M. Principe
SCOTLAND Rab Houston
SEXUALITY Véronique Mottier
SHAKESPEARE Germaine Greer
SIKHISM Eleanor Nesbitt
THE SILK ROAD James A. Millward
SLEEP Steven W. Lockley and Russell G. Foster
SOCIAL AND CULTURAL ANTHROPOLOGY John Monaghan and Peter Just
SOCIALISM Michael Newman
SOCIOLINGUISTICS John Edwards
SOCIOLOGY Steve Bruce
SOCRATES C. C. W. Taylor
THE SOVIET UNION Stephen Lovell
THE SPANISH CIVIL WAR Helen Graham
SPANISH LITERATURE Jo Labanyi
SPINOZA Roger Scruton
SPIRITUALITY Philip Sheldrake
STARS Andrew King
STATISTICS David J. Hand
STEM CELLS Jonathan Slack
STUART BRITAIN John Morrill
SUPERCONDUCTIVITY Stephen Blundell
SYMMETRY Ian Stewart
TEETH Peter S. Ungar
TERRORISM Charles Townshend
THEOLOGY David F. Ford
THOMAS AQUINAS Fergus Kerr
THOUGHT Tim Bayne
TIBETAN BUDDHISM Matthew T. Kapstein
TOCQUEVILLE Harvey C. Mansfield
TRAGEDY Adrian Poole
THE TROJAN WAR Eric H. Cline
TRUST Katherine Hawley
THE TUDORS John Guy
TWENTIETH-CENTURY BRITAIN Kenneth O. Morgan
THE UNITED NATIONS Jussi M. Hanhimäki
THE U.S. CONGRESS Donald A. Ritchie
THE U.S. SUPREME COURT Linda Greenhouse
UTOPIANISM Lyman Tower Sargent
THE VIKINGS Julian Richards
VIRUSES Dorothy H. Crawford
WITCHCRAFT Malcolm Gaskill
WITTGENSTEIN A. C. Grayling
WORK Stephen Fineman
WORLD MUSIC Philip Bohlman
THE WORLD TRADE ORGANIZATION Amrita Narlikar
WRITING AND SCRIPT Andrew Robinson

Lawrence M. Principe

THE SCIENTIFIC REVOLUTION

A Very Short Introduction

Great Clarendon Street, Oxford, OX2 6DP,
United Kingdom

Oxford University Press is a department of the University of Oxford. It furthers the University's objective of excellence in research, scholarship, and education by publishing worldwide. Oxford is a registered trade mark of Oxford University Press in the UK and in certain other countries

First Edition published in 2011

Impression: 7

Published in the United States of America by Oxford University Press
198 Madison Avenue, New York, NY 10016, United States of America

British Library Cataloguing in Publication Data

Data available

ISBN 978–0–19–956741–6

Printed in Great Britain on acid-free paper by
Ashford Colour Press Ltd, Gosport, Hampshire

Contents

Acknowledgements

I would like to thank my friends and colleagues who have read and critiqued all or part of the manuscript of this book, or with whom I have discussed and whined about the challenge of compressing the Scientific Revolution into so compact a form, especially Patrick J. Boner, H. Floris Cohen, K. D. Kuntz, Margaret J. Osler, Gianna Pomata, María Portuondo, Michael Shank, and James Voelkel. I would also like to thank the people who have provided the images for this book: James Voelkel at the Chemical Heritage Foundation; Earle Havens and his colleagues in Special Collections at the Sheridan Libraries, Johns Hopkins University; and David W. Corson and his colleagues at the Rare and Manuscript Collections, Kroch Library, Cornell University.

In particular, I wish to remember here the many conversations held over single-malt Scotch with my colleague and friend Maggie Osler about how to write the history of early modern science. Her premature death leaves the world a poorer and less mischievous place; I dedicate this volume to her memory.

List of illustrations

Introduction

Late in 1664, a brilliant comet appeared in the skies. Spanish observers were the first to note its arrival, but over the following weeks, as it grew in size and brightness, eyes all over Europe turned towards this heavenly spectacle. In Italy, France, Germany, England, the Netherlands, and elsewhere – even in Europe's young colonies and outposts in the Americas and Asia – observers tracked and recorded the comet's motions and changes. Some took careful measurements and argued over calculations of the comet's size and distance, and whether its path through the heavens was curved or straight. Some observed it with the naked eye, others with instruments such as the telescope, an invention then just about sixty years old. Some tried to predict its effects on the Earth, on the weather, on the quality of the air, on human health, and on the affairs of men and the fates of states. Some saw it as an opportunity to test new astronomical ideas, others saw it as a divine portent for good or ill, and many saw it as both. Pamphlets flowed from printing presses, articles and contentions appeared in the new periodicals devoted to natural phenomena, people discussed it in princely courts and academies, in coffee-houses and taverns, while letters full of ideas and data shuttled back and forth among distant observers, weaving webs of communication across political and confessional boundaries. All of Europe watched this spectacle of nature and strove to understand it and to learn from it.

The comet of 1664–5 provides but one instance of the ways in which 17th-century Europeans paid close attention to the natural world around them, interacted with it and with each other. Peering through ever-improving telescopes, they saw immense new worlds – undreamt-of moons around Jupiter, the rings of Saturn, and countless new stars. With the equally new microscope, they saw the delicate details of a bee's stinger, fleas enlarged to the size of dogs, and discovered unimagined swarms of 'little animals' in vinegar, blood, water, and semen. With scalpels, they revealed the internal workings of plants, animals, and themselves; with fire, they analysed natural materials into their chemical components, and combined known substances into new ones. With ships, they sailed to new lands, and brought back amazing reports and samples of novel plants, animals, minerals, and peoples. They devised new systems to explain and organize the world and revived ancient ones, ceaselessly debating the merits of each. They sought for causes, meanings, and messages hidden in the world, for the traces of God's creative and sustaining hand, and for ways to control, improve, and exploit the worlds they encountered with both new technology and hidden ancient knowledge.

The Scientific Revolution – roughly the period from 1500 to 1700 – is the most important and talked-about era in the history of science. Ask ten historians of science about its nature, duration, and impact, and you are likely to get fifteen answers. Some see the Scientific Revolution as a sharp break from the medieval world – a time when we all (Europeans at least) became 'modern'. In this view, the 16th and 17th centuries were truly revolutionary. Others have tried to make the Scientific Revolution into a non-event, a mere illusion of retrospection. More circumspect scholars nowadays, however, recognize the many important continuities between the Middle Ages and the Scientific Revolution, but without denying that the 16th and 17th centuries reworked and built upon their medieval inheritance in significant and stunning ways. Indeed, the 'scientific revolution', now more frequently called the 'early modern period', was a time of *both* continuity and change. It saw a substantial increase in the number of people asking questions about the natural world, a proliferation of new answers to those questions, and the development of new ways of gaining answers. This book describes some of the ways early modern thinkers envisioned and engaged with the worlds around them, what they found in them, and what it all meant for them. It outlines how they laid many of the foundations that continue to undergird modern scientific knowledge and methods, wrestled with questions that continue to trouble us, and even crafted rich worlds of beauty and promise that we have often forgotten how to see.

Chapter 1
New worlds and old worlds

Early modern accomplishments grew upon intellectual and institutional foundations established in the Middle Ages. Many of the questions early moderns strove to answer were posed in the Middle Ages, and many methods used for answering them were products of medieval investigators. Yet early modern scholars loved to disparage the medieval period and to claim that their work was wholly new, despite the fact they retained and relied upon at least as much as they discarded, or retailored it to fit the changing times. Specific changes between the Middle Ages and the early modern period, whether intellectual, technological, social, or political, did not occur simultaneously across Europe. Recognizably 'modern' developments in such areas as medicine, engineering, literature, art, economic and civic affairs were thoroughly established in Italy well before they appeared in more peripheral parts of Europe like England. Similarly, periods of development occurred at different times and speeds within different scientific disciplines. The period roughly 1500 to 1700 – call it what you will – was a rich tapestry of interwoven ideas and currents, a noisy marketplace of competing systems and concepts, a busy laboratory of experimentation in all areas of thought and practice. Text after text from the period testifies to the excitement their authors felt about their own times. One label, one book, one scholar, one generation will not comprehend it in its totality.

To begin to understand it and its significance, we need to look closely at what actually took place then and why.

Understanding the Scientific Revolution requires understanding first its background in the Middle Ages and Renaissance. In particular, the 15th century witnessed significant changes in European society and a massive broadening of Europe's horizons, both literally and figuratively. Four key events or movements fundamentally reshaped the world for people living in the 16th and 17th centuries: the rise of humanism, the invention of movable-type printing, the discovery of the New World, and the reforms of Christianity. While not strictly scientific developments, these changes reshaped the world for thinkers of the period.

The Renaissance and its medieval origins

The term 'Italian Renaissance' usually brings to mind masterpieces of art and architecture by well-known figures like Sandro Botticelli, Piero della Francesca, Leonardo da Vinci, Fra Angelico, and many others. But the Renaissance saw much more than a blossoming of fine arts. Literature, poetry, science, engineering, civic affairs, theology, medicine, and other fields prospered as well. The brilliance and importance of the 15th-century Italian Renaissance for history and for modern culture should not be underestimated. All the same, it should also be remembered that it was not the first significant flowering of European culture after the 5th-century collapse of Classical civilization that followed the fall of the Roman Empire. There had been at least two earlier 'renaissances' (a word which means 'rebirth').

The first, the Carolingian Renaissance, followed the late 8th-century military campaigns of Charlemagne that brought greater stability to Central Europe for much of the 9th century. Charlemagne's court at Aachen (Aix-la-Chapelle) became a centre of learning and culture. The cathedral schools that would later provide the foundations for universities trace their origins to this period. Charlemagne's crowning by Pope Leo III in 800 as 'Emperor of the Romans' encapsulates a basic theme of Carolingian reforms: the attempt to return to the glory of ancient Rome. Architecture, coinage, public works, and even writing styles were devised to reproduce the way imperial Romans had done things, or at least the way 9th-century people imagined the Romans had done things. This flowering was, however, short-lived.

The second 'rebirth' of Latin Europe was much broader and more permanent. Its momentum carried forward, although diminished in intensity, to the start of the Italian Renaissance. This second 'rebirth' was the 'Renaissance of the Twelfth Century', a great explosion of creativity in the sciences, technology, theology, music, art, education, architecture, law, and literature. The triggers for this efflorescence remain open to debate. Some scholars point to a warmer, more favourable climate for Europe beginning in the 11th century (called the 'Medieval Warm Period') coupled with improvements in agriculture that brought enough food and prosperity for Europe's population to double and perhaps triple within a relatively short time. The rise of urban centres, more stable social and political systems, more abundant food, and thus more time for thought and scholarship, all contributed to initiating this Renaissance.

The intellectual appetite of a reawakened Europe found rich fare on which to feed in the Muslim world. As Christian Europe began to push back against the frontiers of Islam in Spain, Sicily, and the Levant, it encountered the wealth of Arabic learning. The Muslim world had become heir to ancient Greek knowledge, translated it into Arabic, and enriched it many times over with new discoveries and ideas. In astronomy, physics, medicine, optics, alchemy, mathematics, and engineering, the *Dār al-Islām* ('Habitation of Islam') towered over the Latin West. Europeans wasted no time in acknowledging this fact, nor in exerting themselves to acquire and assimilate Arabic learning. European scholars embarked upon a great 'translation movement' in the 12th century. Dozens of translators, often monastics, trekked to Arabic libraries, especially in Spain, and churned out Latin versions of hundreds of books. Significantly, the texts they chose to translate were almost entirely in the areas of science, mathematics, medicine, and philosophy.

The Latin Middle Ages had inherited from the Classical world only those texts the Romans possessed; by the end of the empire, only a handful of Roman scholars could read Greek, and therefore virtually the only texts the Romans had to pass on were Latin paraphrases, summaries, and popularizations of Greek learning. It was as if our successors got only newspaper accounts and popularizations of modern science and virtually no scientific journals or texts. Thus scholars of the Latin Middle Ages revered the names of the great authors of antiquity and had descriptions of their ideas, but possessed almost none of their writings.

The 12th-century translators changed all that. They translated works of original Arabic authorship and Arabic translations of ancient Greek works. The majority of ancient Greek texts thus came to Europeans in Arabic dress. From Arabic came the medicine of Galen, the geometry of Euclid, the astronomy of Ptolemy, and virtually the entire corpus of Aristotle we have today – not to mention the more advanced works of Arabic authors in all these fields and more. Around 1200, this explosion of knowledge crystallized into curricula for perhaps the most enduring legacy of the Middle Ages for science and scholarship: the university. Aristotle's writings on natural philosophy formed a core of the curriculum, and his logical works gave rise to Scholasticism, a rigorous and formalized methodology of logical inquiry and debate applicable to any subject, and upon which university studies were based.

The importance of the university as an institutional home for scholarship cannot be overemphasized. As the prominent scholar

Edward Grant writes, the medieval university 'shaped the intellectual life of Western Europe'. While the highest degree in the university was in theology, one could not become a theologian without first mastering the logic, mathematics, and natural philosophy of the day, since those topics were employed routinely in the advanced Christian theology of the Middle Ages. Indeed, most great natural philosophers of the period were doctors of theology: St Albert the Great (now patron saint of natural scientists), Theodoric of Freiburg, Nicole Oresme, Henry of Langenstein. All these figures were educated in, taught in, and found a home in a university.

The vigorous cultural life of the 13th century was checked by the disasters of the 14th. Early in the century – possibly as a result of the end of the Medieval Warm Period – repeated crop failures and famine struck a now overpopulated Europe. At mid-century, the Black Plague swept across Europe with astonishing swiftness, killing its victims within a week of infection. We have no experience today of any loss of life or societal upheaval as rapid, unstoppable, or devastating as the reign of the Black Death. In four years, from 1347 to 1350, it killed roughly half of Europe's population. The first signs of a distinctive Italian Renaissance had begun to appear just before these troubled times – the poet Dante (1265–1321) was active before the plague, while the younger writers Boccaccio (1313–75) and Petrarch (1304–74) lived through it.

Humanism

The Italian Renaissance, fully underway a generation or two after the peak plague years, provided the first key background for the Scientific Revolution: the rise of *humanism*. Humanism proves difficult to define succinctly and rigorously. It is better to speak of *humanisms* – a collection of related intellectual, literary, sociopolitical, artistic, and scientific currents. Among the most widely shared beliefs of humanists was the conviction that they were living in a new era of modernity and novelty, and that this new era was to be measured with respect to the accomplishments of the ancients. They looked for a *renovatio artium et litterarum* (a renewal of arts and letters) to be brought about in part through the study and emulation of ancient Greeks and Romans. Accordingly, it was humanist historians of the Italian Renaissance – such as the Florentines Leonardo Bruni (1369–1444) and Flavio Biondo (1392–1463) – who devised the three-fold periodization of history with which we are all familiar (and from whose implications we still must struggle to free ourselves). According to this periodization, the antiquity of Greece and Rome constitutes the first era, while the third era is that of modernity, beginning of course with the Renaissance authors themselves. Falling between these two high points, according to the humanists, lies a 'middle' period of dullness and stagnation, which is thus called the 'Middle' Ages. Indeed, perhaps the most enduring invention of the Renaissance has been the concept of the Middle Ages, to the extent that we have no name for the period 500 to 1300 that is not suffused with the disdain Italian humanists felt towards it. Given the recent memory of famine and plague years as their immediate background, the restoration of prosperity in Italy around 1400 must surely have seemed the dawn of a 'new age'.

Imitation is supposed to be the sincerest form of flattery, and humanists expressed their admiration of antiquity by imitating Roman styles. Attempts to return to antiquity had happened before, notably in the Carolingian Renaissance 600 years earlier. The grandeur of Rome casts a very long shadow indeed in human memory. The humanist hunger to know more about that past era expressed itself in a quest for long-lost Classical texts. One early humanist, Poggio Bracciolini (1380–1459), taking advantage of recesses during the reform-minded Church Council of Konstanz (1414–18), where he was employed as apostolic secretary, ransacked nearby monastic libraries searching for survivals of Classical literature. He found Quintilian on rhetoric and previously unknown orations of Cicero, but – of greater importance for the history of science – he found also Lucretius' *On the Nature of Things*, a work that presented ancient notions of atomism, Manilius on astronomy, Vitruvius on architecture and engineering, Frontinus on aqueducts and hydraulics. These works had been copied and preserved through the centuries by medieval monks, and had lain – perhaps in just a single surviving copy – in their monastic libraries for generations.

The humanists' recovery of Roman learning was paired with a revival of the study of Greek. The background for the revival of Classical Greek, almost completely unstudied in the Latin West for a thousand years, was the arrival of Greek diplomats and churchmen on embassies to Italy around 1400. Their mission was to secure aid against the Turkish threat and a reunion of Eastern and Western Churches, divided by schism since 1054. One of the first, Manuel Chrysoloras (c. 1355–1415), arrived as a diplomat but stayed as a teacher of Greek; many prominent humanists were his students. Their appetites whetted for Greek texts, Italians travelled to Constantinople to hunt after manuscripts. Guarino da Verona (1374–1460) brought back crates of manuscripts that included Strabo's *Geography*, which he then translated. It is said that one crate of manuscripts was lost in transit, which made Guarino's hair turn grey overnight from grief. The Greek delegation to the Council of Florence in the 1430s included two notable Greek scholars. One was Basilios Bessarion (1403–72), later made a cardinal, who gave his collection of nearly a thousand Greek manuscripts to Venice. The other was a strange character named Georgios Gemistos, known as Pletho (c. 1355–c. 1453), who later advocated a return to ancient Greek polytheism. Pletho taught Greek in Florence, and brought the works of Plato and Platonists to the attention of the West. His teaching led the ruling Cosimo I de' Medici to found a Platonic Academy in Florence. Its first leader, Marsilio Ficino (1433–99), translated the works of Plato and texts by several later Platonists, most of which had been unknown to Western European readers.

Thus the 15th century saw the recovery of huge numbers of ancient texts – many on scientific and technological topics – much as the 12th century had done. But humanists were distinguished not so much by a love of texts, as by a love of *pure and accurate* texts. They disdained the texts of Aristotle and Galen used in universities as corrupt – full of barbarisms, 'Arabisms', accretions, and errors. They rejected Scholasticism as sterile, barbarous, and inelegant. They considered the universities (particularly the northern ones, less so those in Italy) as relics of those stagnant 'Middle' Ages, and chided their scholars for writing a degraded Latin, devoid of elegance. Thus an important feature of humanism was its establishment of new scholarly communities outside the universities.

There is a modern misconception that humanists were somehow secularist, irreligious, or even anti-religious. It is true that some humanists criticized ecclesiastical abuses and disdained Scholastic theology, but in no way whatsoever did they reject Christianity or religion. Indeed, many advocated church reforms parallel with their desired reform of language – by a return to antiquity, to the Church of the first several centuries AD. Many humanists were in Holy Orders, employed in ecclesiastical administration, or supported by church benefices, and the Catholic hierarchy patronized humanism. Many Renaissance-era Popes were fervent humanists – particularly Nicholas V, Sixtus IV, and Pius II – as were their cardinals and courts, where humanists were encouraged. The modern error comes from a confusion with so-called *secular humanism*, an invention of the 20th century that has no counterpart in the early modern period.

Renaissance humanism's impact on the history of science and technology was both positive and negative. On the positive side, humanists made available hundreds of important new texts, and promoted a new level of textual criticism. The reintroduction of Plato, thanks especially to his adoption of Pythagorean mathematics, raised the status of mathematics and provided an alternative to the Aristotelianism favoured at universities. The desire to measure up to the ancients inspired engineering and building projects across Italy, with the ancient engineers Archimedes, Hero, Vitruvius, and Frontinus as models. On the downside, the adulation of antiquity could go too far by rejecting everything after the fall of Rome as barbarism. It is thus that Europe began to lose its respect for and knowledge of Arabic and medieval achievements, which in the sciences, mathematics, and engineering were – let there be no doubt – substantial advancements over the ancient world.

The invention of printing

The invention of movable-type printing around 1450 well served the humanist interest in texts. This invention, or at least its successful deployment, is credited to Johannes Gutenberg (c. 1398–1468), originally a goldsmith in Mainz. The key to movable-type printing was the creation of cast metal type, each bearing a single raised letter. These type could be assembled into full pages of text, their surfaces smeared with an oil-based ink and pressed against paper, thus printing an entire page (or set of pages) at once. After printing numerous copies, the page of type could be taken apart and the letters readily rearranged into the next set of pages. Previously, books had to be copied by hand, resulting in slow production and high price. The late medieval growth of universities and increase in literacy created a demand for books that outstripped the supply, exerting pressure to produce books more quickly, thus leading to book-making enterprises outside the traditional monastic and university scriptoria. This increased production led to more copying errors – something humanists deplored. Printing allowed for faster and more reliable production, although the labour involved in paper-making, typesetting, and printing meant that books remained expensive. (Gutenberg's Bible, printed in 1455, cost 30 florins, more than a year's salary for a skilled workman.)

The transition to print was not immediate. Manuscripts continued to exist alongside books, although their use was increasingly limited to the restricted circulation of private, rare, or privileged materials. Printed typefaces mimicked manuscript writing; in Northern Europe this meant Gothic bookhands, but Italy, Venice in particular, soon became the centre of the printing industry. Italian printers, such as Teobaldo Mannucci, better known by his Latinized humanist name Aldus Manutius (1449–1515), adopted the cleaner, crisper shapes of letters developed by Italian humanists (which they thought imitated the way Romans wrote), thereby creating fonts that not only displaced older ones, but also formed the basis for most fonts used today; hence our elegant slanted font is still known as 'Italic'.

Printing presses sprang up rapidly across Europe. By 1500, there were about a thousand in operation, and between thirty and forty thousand titles had been printed, representing roughly ten million books. This flood of printed material only increased throughout the 16th and 17th centuries. Books became steadily less expensive (often with a loss of quality) and easier for less wealthy buyers to obtain. Printing allowed for faster communication through broadsides, newsletters, pamphlets, periodicals, and a slew of other paper ephemera. Although most of these ephemera perished soon after their production (like last week's newspaper), such items were very common in the early modern period. The press thus created a new world of the printed word – and of literacy – like never before known.

One easily overlooked feature of printing was its ability to reproduce *images and diagrams*. Illustrations posed a problem for the manuscript tradition since the ability to render drawings accurately depended upon the copyist's draftsmanship, and often upon his understanding of the text. Consequently, every copy meant degradation for anatomical renderings, botanical and zoological illustrations, maps, charts, and mathematical or technological diagrams. Some copyists simply omitted difficult graphics. Printing meant that an author could oversee the production of a master woodcut or engraving, which could then produce identical copies easily and reliably. Under such conditions, authors were more willing and able to include images in their texts, enabling the growth of scientific illustration for the first time.

Voyages of discovery

Since a picture is worth a thousand words, the ability to illustrate proved especially important given the strange new reports and objects that would soon flood Europe. This information came from new lands being contacted directly by Europeans. The first source was Asia and sub-Saharan Africa. European contact with these places came about thanks to Portuguese attempts to open a sea route for trade with India in order to cut out the middlemen – predominantly Venetians and Arabs – who controlled the overland and Mediterranean routes. In the early 15th century, the Portuguese prince known as Henry the Navigator (1394–1460) began sending expeditions down the west African coast, establishing direct contact with traders in sub-Saharan Africa. Portuguese sailors pushed on further and further south, eventually rounding the Cape of Good Hope in 1488, and culminating in Vasco da Gama's successful trading voyage to India in 1497–98. The Portuguese established trading outposts all along the route, many of which remained Portuguese possessions until the middle of the 20th century, and eventually extended their regular voyages as far as China, transporting luxury goods like spices, precious stones, gold, and porcelain back to Europe. They also brought back stories of distant lands, strange creatures, and unknown peoples.

This broadening of European horizons did not begin abruptly in the Renaissance. The Middle Ages laid the foundations for Renaissance-era voyages. Indeed, the eastward voyages of the 15th century re-established contacts that had been made in the

13th but cut off in the 14th due to political upheavals in Asia. Medieval travellers, often members of the two new religious orders of the 13th century – Dominicans and Franciscans – embarked on distant religious and ambassadorial missions to an extent we are only now beginning to recognize. They established religious houses across Asia all the way to Peking, as well as in Persia and India, and sent back information to Europe that informed and inspired later mercantile voyages. These medieval travels resulted in a broader sense of the place of Europe within a much larger world to be explored.

While the Portuguese were opening sea routes eastward towards Asia, Christopher Columbus was staring off in the opposite direction. Convinced that the circumference of the earth was about one-third less than the fairly accurate estimates made in antiquity and still widely known in Europe, Columbus imagined that he could reach East Asia faster by sailing westwards. This mistaken impression was in part due to Ptolemy, the 2nd-century geographer and astronomer. Humanists had recently recovered his *Geography*, which included an anomalously small figure for the size of the earth and considerably overestimated the eastward extent of Asia. Financial backers of Columbus were duly sceptical; they recognized that the westward route was the longer way around, and without intermediate places to take on fresh supplies, the crew would starve. (*No one* thought Columbus would 'sail off the edge of the earth', since the sphericity of the Earth had been fully established in Europe for over 1,500 years before Columbus. The notion that people before Columbus thought that the Earth was flat is a 19th-century invention. Medievals would have had a good laugh at the idea!) Hence, when in 1492 Columbus's ships struck land in the Caribbean, he thought he had reached Asia rather than discovered a new continent.

Whether or not Columbus later acknowledged his mistake, others quickly did, and hastened to travel to this New World. News of the new continent spread quickly, aided by the young printing press, and in 1507, a German cartographer gave the new lands a name – America – after the Italian explorer Amerigo Vespucci. Thanks to these maps and Vespucci's accounts of South America published with them, the name stuck. In 1508, King Ferdinando II of Spain created the position of chief navigator for the New World for Vespucci. This new position existed within the Casa de Contratación (House of Trade), a centralized bureau founded in 1503 not only for collecting taxes on goods brought back to Spain, but also for collecting and cataloguing information of all kinds from returning travellers, for training pilots and navigators, and for constantly updating master maps with new information gleaned from every returning ship's captain. The knowledge and practical know-how collected in Seville helped Spain establish the first empire in history upon which 'the sun never set'.

Other nations, not wishing to be left out of the territories and wealth Spain and Portugal were amassing, joined the fray, although trailing the Iberians by a century or more. Thus for a hundred years, virtually all the New World reports and samples that transformed European knowledge of plants, animals, and geography came into Europe through Spain and Portugal. It is hard to imagine the flood of data that poured into Europe from the New World. New plants, new animals, new minerals, new medicines, and reports of new peoples, languages, ideas, observations, and phenomena overwhelmed the Old World's ability to digest them. This was true 'information overload', and it demanded revisions to ideas about the natural world and new methods for organizing knowledge. Traditional systems of classifying plants and animals were exploded by the discovery of new and bizarre creatures. Observations of human habitation virtually everywhere explorers could reach refuted the ancient notion that the world was divided into five climatic regions – two temperate ones and three rendered uninhabitable due to excessive heat or cold. Exploiting the enormous economic potential of the Americas and Asia required fresh scientific and technological skills. Geographical data and the recording of sea routes drove the creation of new mapping techniques, while getting safely and reliably between Europe and the new lands demanded improvements to navigation, shipbuilding, and armaments.

Reforms of Christianity

While voyages around the world exposed Europeans to a diversity of religious perspectives, such perspectives were also diversifying at home. The year 1517 marks the beginning of a deep, often violent, and continuing rupture within Christianity. In that year, the Augustinian priest and theology professor Martin Luther (1483–1546) proposed his famous 'Ninety-Five Theses' in the university town of Wittenberg. These theses, or propositions, were written in the format of topics for Scholastic disputation, and centred on inappropriate and theologically indefensible contemporaneous local practices involving the sale of indulgences. While similar debates over practical and doctrinal issues were common fare in the disputative university culture of the Middle Ages, Luther's protest passed beyond the usual confines of scholarly theological disputation and quickly became a broad-based political and social movement out of Luther's control. Although initially quite mild, Luther's claims became increasingly bold and confrontational, moving from relatively minor issues of local practices into serious doctrinal matters. These claims were quickly disseminated by the printing press, deepened by linkages to local nationalism, and abetted by Germanic rulers who saw separation from Rome as favourable to their political interests. A local protestation thus unexpectedly became Protestantism. Protestantism almost immediately splintered into sparring sects. Catholic-Lutheran controversies were soon joined by Lutheran-Calvinist ones, then by intra-Calvinist ones, and so on. The so-called 'Wars of Religion' – often motivated more by political and dynastic manoeuvres than by doctrinal issues – convulsed Europe, particularly Germany, France, and England, for the next century and half.

Luther himself was no humanist, although some of his notions, such as an emphasis on a literal reading of the Bible as opposed to the allegorical readings favoured by Catholics, bear resemblances to humanist emphases on texts. But these resemblances are outweighed by his suspicion of Classical ('pagan') literature and ideas and his desire to expunge books from the Bible (such as the Letter of James) that disagreed with his personal notions. The much more learned Philipp Melanchthon (1497–1560), however, was quite a different story. Melanchthon's very name testifies to his humanism, translated into Classical Greek from the original barbarous German Schwartzerd ('black earth'). His great uncle, Johannes Reuchlin, who suggested this 'self-classicization', was the most prominent humanist in Germany. In the wake of the Lutheran rejection of university Scholasticism, Melanchthon (who as a humanist also disliked Scholasticism) renovated university curricula and pedagogy in German universities – in particular, Luther's own University of Wittenberg – as they converted from Catholic to Lutheran. The new curricula he devised earned him the title *Praeceptor Germaniae* ('Teacher of Germany'). His approach was not to banish Aristotle, but rather – in true humanist fashion – to banish medieval 'accretions' to Aristotle and to use better editions of the Greek philosopher. New Protestant universities found themselves in the enviable position of having to start afresh, that is, with a reduced burden of established methods, and were thus able to incorporate new subjects and approaches that had not found a place in older institutions.

Within Catholicism, reform movements were also underway. In the 15th century, church councils addressed some issues, although not very successfully. More dramatic was the Council of Trent (1545–63), an Ecumenical Council convened to respond to Protestantism by addressing corruption, clarifying doctrines, standardizing practices, and centralizing disciplinary oversight. The Council of Trent, the most important post-medieval church council until Vatican II (1962–5), launched the Catholic Reform, or 'Counter-Reformation'. Its measures included improved education for priests, a reform many humanists had been advocating, but also increased oversight of orthodoxy including in published works. Tridentine reforms were taken up most avidly by a newly organized society of priests, the Society of Jesus, or Jesuits. Organized by St Ignatius Loyola and given papal authorization in 1540, the Jesuits devoted themselves especially to education and scholarship, and made significant contributions specifically to science, mathematics, and technology.

The broader impact of the Jesuits, besides preaching for a return of Protestants to Catholicism, lay in the hundreds of schools and colleges they established within the first years of their existence. Jesuit pedagogy rested upon an innovative style of teaching and curriculum, one that preserved the importance of Aristotelian methods, but paired that with new emphasis on mathematics (by 1700, more than half of all the professorships of mathematics in Europe were held by Jesuits) and the sciences. Jesuit schools were often the first to teach some of the new scientific ideas of the Scientific Revolution, and educated many of the thinkers responsible for them. Jesuits spread out across the globe along the newly opened trade routes, establishing a high-profile presence (and schools, of course) in China, India, and the Americas, and the first global correspondence network. This network channelled everything from biological specimens and astronomical observations to cultural artefacts and extensive reports of native knowledge and customs back to Rome. The Jesuit attitude in studies of science and mathematics expresses their motto 'to find God in all things'. While Jesuits emphasized this incentive, it was not unique to them – it undergirded virtually the entire Scientific Revolution.

The new world of the 1500s

Europeans of the 16th century inhabited a new and rapidly changing world. As in our own fast-paced days, many saw this situation as a source of anxiety, while others saw a world of opportunities and possibilities. The horizons of Europe had been expanded in every sense. Europeans had rediscovered their own past, encountered a wider physical and human world, and created new approaches and fresh interpretations of older ideas. Indeed, the best image for their world would be that of a tumultuous and richly stocked market place. A cacophony of voices promoted a diversity of ideas, goods, and possibilities. Throngs jostled elbows to test, purchase, reject, praise, criticize, or just touch the varied merchandise. Almost everything was up for grabs. Whether we conclude the 'Scientific Revolution' to be something entirely new, or a revival of the intellectual ferment of the late Middle Ages after the interruption of the baleful 14th century, there can be no doubt that the learned inhabitants of the 16th and 17th centuries saw their time as one of change and novelty. These were exciting times; times of new worlds indeed.

Chapter 2
The connected world

When early modern thinkers looked out on the world, they saw a *cosmos* in the true Greek sense of that word, that is, a well-ordered and arranged whole. They saw the various components of the physical universe tightly interwoven with one another, and joined intimately to human beings and to God. Their world was woven together in a complex web of connections and interdependencies, its every corner filled with purpose and rich with meaning. Thus, for them, studying the world meant not only uncovering and cataloguing facts about its contents, but also revealing its hidden design and silent messages. This perspective contrasts with that of modern scientists, whose increasing specialization reduces their focus to narrow topics of study and objects in isolation, whose methods emphasize dissecting rather than synthesizing approaches, and whose chosen outlooks actively discourage questions of meaning and purpose. Modern approaches have succeeded in revealing vast amounts of knowledge about the physical world, but have also produced a disjointed, fragmented world that can leave human beings feeling alienated and orphaned from the universe. Virtually all early modern natural philosophers operated with a wider, more all-embracing vision of the world, and their motives, questions, and practices flowed from that vision. We have to understand their worldview if we are to understand their motivations and methods in investigating that world.

The concept of a tightly connected and purposeful world derives from many sources, but above all from the two inescapable giants of antiquity, Plato and Aristotle, and from Christian theology. From Platonic sources, particularly the thinkers called Late Platonists or Neoplatonists – philosophers actively developing Plato's ideas in Hellenized Egypt during the first centuries of the Christian Era – comes the idea of a *scala naturae*, or ladder of nature. According to this conception, everything in the world has a special place in a continuous hierarchy. At the very top is the One – the utterly transcendent, eternal God, from whom everything else derives existence. The One emanates creative power that brings everything else into existence. The further this power radiates from its Source, the lower and more unlike the One are the things it creates. At the bottom lies inert, lifeless matter. The rungs in between, in ascending order, are filled with vegetable and animal life, then human beings, and then spiritual beings such as *daimons* and lesser gods. The goal of some Neoplatonists was to climb the ladder as it were, to became more spiritual and less material, to free the human soul – our most noble part – from the blindness caused by its descent into matter, and to rise through the levels of spiritual beings in journey towards the One. This late antique conception both influenced and was influenced by Christian doctrines, and could be readily adapted to orthodox Christian beliefs by replacing the pagan *daimons* and lesser gods with orders of angels, and the One with the Christian God, as was suggested by the 5th-century Christian Neoplatonist pseudo-Dionysius the Areopagite. Thanks to such Christianization, the idea of the *scala naturae* remained well known throughout the Latin Middle Ages, even if the ancient Platonic texts upon which it was based were lost for centuries.

These Platonic texts were among those rediscovered by humanists in the Renaissance and translated by Marsilio Ficino. Ficino also acquired, translated, and published a set of texts attached to the name Hermes Trismegestus, meaning Hermes 'the Thrice-Great', a supposed ancient Egyptian sage contemporary with Moses.

What Ficino obtained was a small selection out of a huge mass of diverse *Hermetica* (writings attributed to Hermes) dating from about the 3rd century BC to the 7th AD. Although initially believed to be much older, Ficino's *Hermetica* probably dates from the 2nd and 3rd centuries AD. Its importance lies in its Neoplatonic character that emphasizes the power of human beings, their place in the connected world of the *scala*, and their ability to ascend it. Many Renaissance readers found what they thought to be foreshadowings of Christianity in the *Hermetica*, and thus Hermes Trismegistus took on the status of a pagan prophet, and accordingly he can be found depicted among the prophets in the cathedral of Siena.

The *scala* envisions of a world in which every creature has a place, and each creature is linked to those immediately above and below it, such that there is a gradual and continuous rise from the lowest level to the highest, without gaps, along what has been called 'the Great Chain of Being'. A related concept – present in the *Timaeus*, Plato's account of the origin of the universe, and the only work of Plato known to the Latin Middle Ages – is that of the *macrocosm* and *microcosm*. These two Greek words mean, respectively, the 'large ordered world' and the 'little ordered world'. The macrocosm is the body of the universe, that is, the astronomical world of stars and planets, while the microcosm is the body of the human being. The essential idea is that these two worlds are constructed on analogous principles, and so bear a close relationship to each other. A late contribution to the *Hermetica*, an 8th-century Arabic work called the *Emerald Tablet*, concisely summarizes this view in a terse motto well known in early modern Europe: 'as above, so below'. For Plato, the linkage of man's microcosm with the planetary macrocosm had a practical moral meaning – we should look to the orderly, rational workings of the heavens as a guide for governing ourselves in an orderly, rational way. For early modern Europeans, the microcosm–macrocosm linkage had, above all, a medical meaning – it undergirded medical astrology. The various planets have particular effects upon particular human organs, whereby they can influence the bodily functions (see Chapter 5).

A second major contributor to the view of an interconnected and purposeful world comes from Aristotelian ideas about how to gain knowledge. According to Aristotle, proper knowledge of a thing is 'causal knowledge'. That term requires explanation. Aristotle argued that knowing a thing requires identifying its four 'causes', or reasons for existing. The first of these, the *efficient cause*, describes what or who made the thing. The *material cause* describes what the thing is made of. The *formal cause* tells what physical characteristics make the thing what it is, in other words, an inventory of its qualities. The most important cause for Aristotelians, and the most difficult one for moderns to get their minds around, is the *final cause*. The final cause tells what the thing is for, that is, what its goal in existing is, and for Aristotle, everything has a goal or purpose. These 'causes' can be illustrated using a statue of Achilles. The statue's efficient cause is the sculptor, its material cause is marble, its formal cause is the beautiful body of Achilles, and its final cause is to celebrate the memory of Achilles. There can be more than one of each of the causes (for example, the statue might also have the final cause of being decorative, or perhaps, in some Attic house, to act as a coat rack).

The crucial point is that Aristotelian forms of knowledge, particularly in regard to the efficient and final causes, acted to define objects *in the context of their relationship to other objects*. Coming to know a thing meant being able to position it within a network of relationships with other things, particularly the things that bring it into being and that make use of it. In the Christian context of Europe, the final cause harmonized well with the idea of divine design and providence. Final causes in nature were part of God's plan for creation, implanted and encoded within created things by the First Efficient Cause.

Writers of the early modern period expressed their understanding of a connected world in many different ways. The English natural philosopher Robert Boyle (1627–91), renowned for his work in chemistry (chemistry students still have to learn Boyle's Law that the volume of a gas is inversely proportional to the pressure exerted upon it), wrote that the world is like 'a well contriv'd Romance'. Here, Boyle alludes to the massive French novels of his day (of which he was very fond). These romances often run to more than two thousand pages in length, and feature a memory-taxing myriad of characters whose complex storylines constantly converge and diverge in surprising ways, full of revelations about who is secretly in love with whom and who is really whose long-lost brother, child, or what-not. For Boyle, the Creator is the ultimate romance writer, and scientific investigators are the readers trying to figure out all the relationships and crisscrossing storylines in the world He wrote.

The Jesuit polymath Athanasius Kircher (1601/2–80), who maintained a museum of wonders in Rome and was a centre of Jesuit correspondence about natural philosophy, portrayed the connected world in an elegant Baroque frontispiece to his encyclopaedic work on magnetism (Figure 1).

The image shows a series of circular seals, each bearing the name of one branch of knowledge: physics, poetry, astronomy, medicine, music, optics, geography, and so on, with theology at the top. A single chain connects the seals together, expressing the inherent unity of all branches of knowledge. For early moderns, there were no strict barriers that kept sciences, humanities, and theology insulated from one another – they formed interlocking ways of exploring and understanding the world. In Kircher's image, these branches of knowledge stand chained to three larger seals representing the three chief parts of the natural world: the siderial world (everything farther away than the Moon), the sublunar world (the Earth and its atmosphere), and the microcosm (human beings). These three parts of the world are likewise

1. Engraved title page to Athanasius Kircher, *Magnes sive de magnetica arte* (Rome, 1641) expressing the interconnectedness of the branches of knowledge and of God, humanity, and nature

chained together indicating the inescapable interdependence that exists between them. At the centre of the entire image, in direct contact with each one of the three worlds equally, stands the *mundus archetypus* – the archetypal world, that is, the mind of God that not only created everything, but also contains within itself the models or archetypes of everything possible in the universe. Kircher completes his image with the Latin motto: 'Everything rests placidly, connected by hidden knots.'

This sense of connectedness both between disciplines and between various facets of the universe characterizes *natural philosophy* – the discipline practised by early modern students of the natural world. Natural philosophy is closely related to what we familiarly call *science* today, but is broader in scope and intent. The natural philosopher of the Middle Ages or of the Scientific Revolution studied the natural world – as modern scientists do – but did so within a wider vision that included theology and metaphysics. The three components of God, man, and nature were never insulated from one another. Natural philosophical outlooks gradually gave way to more specialized and narrow 'scientific' ones only during the 19th century (the age in which the word 'scientist' was first coined). The work and motivations of early modern natural philosophers cannot be properly understood or appreciated without keeping the distinct character of natural philosophy in mind. Their questions and goals were not necessarily our questions and goals, even when the very same natural objects were being studied. Hence, the history of science cannot be written by pulling scientific 'firsts' out of their historical context, but only by seeing with eyes and minds of our historical characters.

Natural 'magic'

The 'cosmic' perspective was widely shared in the 16th and 17th centuries, and it undergirded a variety of practices and projects, even if different thinkers considered the interconnections in the world to be of varying degrees of importance to their work. The facet of natural philosophy most closely tied to this vision of the world was *magia naturalis*. It is misleading to translate this Latin term directly into English as 'natural magic'. The word 'magic' naturally makes modern readers think of costumed men pulling rabbits out of hats, or of wizened black-robed characters in pointy hats mumbling over cauldrons, or, rather more benignly, of Harry Potter and Hogwarts. The *magia naturalis* of the early modern period was, however, something very different; it forms an important part of the history of science.

Magia is perhaps best translated for moderns as 'mastery'. The goal of the practitioner of *magia*, called a *magus*, is to learn and to control the connections embedded in the world in order to manipulate them for practical ends. Look again at Kircher's frontispiece. In the upper left-hand corner, *magia naturalis* is listed among the branches of knowledge, between arithmetic and medicine. Kircher symbolizes it with the turning of a sunflower to follow the Sun across the sky throughout the day. (Several plants display this behaviour, known as *heliotropism*.) Why does the sunflower always turn towards the Sun while most plants do not? Clearly, there must be some special link between Sun and sunflower. The ability of the sunflower to follow the Sun provided a prime example of the hidden connections and forces in the world that the magus endeavoured to identify and control.

Medieval Aristotelians divided properties of a thing into two groups. The first were *manifest qualities* – qualities that anybody endowed with sense organs could detect. Hot, cold, wet, and dry were the primary qualities. Other qualities included things like smooth, rough, yellow, white, bitter, salty, sonorous, fragrant, and so forth – all things that activated the senses. After all, Aristotelianism was fundamentally a common-sense way of engaging with the world. Aristotelians used these manifest qualities to explain the action of one thing upon another: cooling drinks lower a fever because cold counteracts hot, for example. But some objects acted in weird ways that manifest qualities could not explain. These objects were held to have *hidden qualities* (*qualitates occultae*, often misleadingly translated as 'occult qualities') that we cannot detect with our senses. These qualities often acted in highly specific ways, suggesting a special, invisible connection between specific things and the objects they acted upon. Medieval natural philosophers compiled lists of such phenomena. One classic example is the magnet. We can sense nothing about the lodestone (a naturally magnetic mineral) that could possibly explain its mysterious ability to attract iron specifically. The same is true of the apparent attraction between the Sun and the sunflower, the turning of a compass needle towards the pole star, the sleep-inducing effect of opium, the Moon's effect on the tides, and many other things. *Magia naturalis* was the endeavour to seek out these hidden qualities of things and their effects, and to make use of them.

How did one go about finding these connections, these 'hidden knots', in nature? One way was to observe the world closely. Everyone can agree that careful observation is a crucial starting point for scientific investigation; the pursuit of *magia naturalis* promoted such observation. A method of equal importance lay in mining the records of earlier observers of nature – accounts and observations, ranging from the commonplace to the bizarre, recorded in various texts from contemporaneous times back to the ancient world. Much *magia* was therefore based on a careful reading of texts in humanist fashion, building up complex networks by compiling claims from earlier writers. Given the immense variety of nature, the task of the aspiring *magus* is mind-bogglingly immense – no less than cataloguing the properties of everything. Could there be a shortcut? Some natural philosophers believed that nature contained clues to guide the magus, perhaps as hints implanted there by a merciful God who wants us to understand His creation and benefit from it. The *doctrine of signatures* claims that some natural objects are 'signed' with indications of their hidden qualities. Often, this means that two connected objects look somehow similar, or have some analogous characteristics; for example, the sunflower not only follows the Sun, its blossom actually *resembles* the Sun in colour and shape. Various parts of plants resemble various parts of the human body; a walnut nestled in its shell looks remarkably like a brain inside the skull. Is this a sign that walnuts would provide good medicines for the brain? The practitioner of *magia* would have to try these things out to be sure, but observation coupled with the idea of signatures provided a useful point of departure for investigating, explaining, and using the natural world.

The doctrine of signatures represents but one facet of a broader mode of analogical thinking ubiquitous in the early modern period. While moderns would tend to see such similarities as mere coincidence or accident, or as 'poetic' rather than physical, many early moderns saw things quite differently – they *expected* analogical links between different parts of the world, and the discovery of an analogy or symmetry in nature signified for them a real connection between things. Rather than being the product of human imagination, every analogy between two objects in the natural world marked out another line in the blueprint of creation, a visible sign of a hidden connection divinely implanted in the universe. Thus, arguments from analogy carried special strength and evidentiary power beyond what we are accustomed to give them today. The sureness of this linkage was founded upon an unshakable faith in a cosmos that was not random or fortuitous, but rather one that was suffused with meaning and purpose, guided in various ways by divine wisdom and providence for the benefit of human beings. This certainty, and the attendant use of analogical reasoning, was not the exclusive property of those interested in *magia naturalis*, but of virtually *every* serious thinker of the period.

Using direct observation, analogy, textual authorities, and signatures, early modern thinkers compiled huge aggregates of things they considered to be linked. For example, what else might relate to the Sun–sunflower connection? The Sun is the source of warmth and life in the macrocosm, its counterpart in the microcosm must be the heart. (Have yet another look at Kircher's frontispiece – there is a tiny Sun in the place of the heart in the human figure representing the microcosm.) The Sun is the most noble of the heavenly bodies, brilliant and yellow, and thus it bears a similitude to gold in the mineral realm, and further afield to all yellow or golden things. In the animal realm, the Sun causes the rooster to crow, indicating a special link between the two. The lion, with its tawny colour, royal status, and head that resembles the Sun (its mane frames its head like solar rays), also seems linked to the Sun. Likewise, the bravery of the lion corresponds in turn with the heart. Sun, sunflower, heart, gold, yellow, rooster, and lion all bear links of commonality and thus real but hidden connections. For the advocates of *magia naturalis*, these analogical links translate into operative links that can be put to use. The most down-to-earth application would involve using gold or sunflowers to make a medicine for the heart – but things could get much more dramatic, as we shall see.

Opinions varied as to what actually linked objects bound up in these webs of correspondence, but they were usually considered to function by means of 'sympathy', which literally means 'suffering together or receiving action together'. Think of two well-tuned lutes on opposite sides of a room, pluck a string on one of them, and the corresponding string on the other will immediately start to vibrate and hum on its own, echoing the note plucked on the first lute. Today, we still call this phenomenon *sympathetic* vibration. For early modern thinkers, this phenomenon exemplified the operation of unseen links acting at a distance between things that were 'in tune' with one another. Some argued that a medium was necessary to transmit the action between spatially separated objects; Aristotle had argued that one thing could not act on another thing a distance away without an intervening medium to carry the effects. In the case of lute strings, for example, we know that the intervening air carries the vibrations between the two instruments. For other sympathetic actions, this medium might be the so-called *spiritus mundi*, or spirit of the world – a universal, all-penetrating incorporeal or quasi-corporeal substance, capable of keeping even distant objects in virtual contact with one another by transmitting influences from one to the other. This 'spirit' was not some sentient supernatural entity; rather, it is the macrocosmic equivalent of the microcosmic animal spirits, the subtle substance in our bodies that transmits the command 'move!' through the nerves to our feet when our intellect realizes that a two-ton truck is speeding towards us. The spirit of the world likewise carries 'signals' from the Sun to the sunflower or from the Moon to the waters of the sea. Once again, the microcosm and the macrocosm are reflections of one another; both contain spirits that transmit signals. Incidentally, this analogous nature should also mean that the macrocosm itself has a soul of some sort – a point Plato asserts in the *Timaeus* and is especially difficult for moderns to understand – the next chapter returns to this point.

Practical 'mastery' from the kitchen to the study

The theory of natural magic in regard to a connected world is impressive, even elegant and beautiful, yet the key feature of *magia naturalis* is practical application. The practical parts of early modern *magia* range from the banal to the sublime, the former often having little to do with any theoretical foundations. The book *Magia naturalis* of Giambattista della Porta (1535–1615) provides a good example. Della Porta is renowned for establishing in Naples the earliest scientific society – the Academy of Secrets – and for being a member of the Accademia dei Lincei, the early 17th-century scientific society that counted Galileo as a member. The first chapter of Della Porta's book recapitulates the principles of an interconnected world, noting how magic 'is the survey of the whole course of nature' and 'the practical part of natural philosophy'. He advises his reader to 'be prodigal in seeking things out; and while he is busy and careful in seeking, he must be patient also . . . neither must he spare any pains: for the secrets of nature are not revealed to lazy and idle persons'. The practical secrets of nature that the rest of della Porta's book reveals do include observations about magnetism and optics, but the majority of the book is a miscellany of recipes for everything from making artificial gems and fireworks, to animal and plant breeding, to household hints about making perfumes, roasting meat, and preserving fruit, none of which draws upon any theoretical conception of the world. Della Porta's book fits instead with a tradition of 'books of secrets' that became increasingly popular throughout the 16th and 17th centuries, some of which were reprinted even into the 19th. Many such books begin with an exposition of grand and lofty notions about the cosmos, but consist principally of recipes for household management or cottage industries, and contain little or nothing about the nature of the world.

At the sublime end of the scale stands Marsilio Ficino (1433–99), whose practical application of the connectedness of the world was expressed in ways of living and in rituals. Ficino often complained of his melancholy temperament; perhaps he suffered from what we now label as depression. The established medicine of the day held that a preponderance of black bile – one of the four 'humours' of the body that must remain in balance to provide health – produces depression. Indeed, the Greek term for black bile – *melaina cholē* – is the origin for our word *melancholy*. (In the same way, personalities that are still called sanguine, choleric, and phlegmatic arise from the preponderance of one of the other three bodily humours: blood, yellow bile, or phlegm, respectively; see Chapter 5.) Ficino explored the connection between the scholarly life and melancholy, and proposed lifestyle changes for his fellow intellectuals to help them address the problem. He formulated a diet and medicinal supplements to prevent the formation of excess black bile in the body, and his 'On Obtaining Life from the Heavens' proposes using celestial influences to counteract this occupational hazard of scholars.

Physicians considered black bile to have the manifest qualities of cold and dry. The planet Saturn shares these qualities, and thus the two bear a sympathetic connection. Therefore, anything in the web of correspondences with black bile and Saturn was to be avoided. The opposing qualities of the Sun (hot–dry) and Jupiter (hot–wet) counteract the cold–dry of black bile, and so by analogical extension anything in the web of correspondences with the Sun and Jupiter could help counteract scholarly melancholy. (Our word 'jovial' literally means 'relating to Jupiter', an indication preserved in our language of how thoroughly entrenched and accepted this reasoning really was.) Thus, in order to make use of sympathetic links to the Sun, the Florentine humanist suggested wearing yellow and golden clothes, decorating one's chamber with heliotropic flowers, getting lots of sunlight, wearing gold and rubies, eating 'solar' foods and spices (like saffron and cinnamon), hearing and singing harmonious and stately music, burning myrrh and frankincense, and drinking wine in moderation. For some readers, however, he did tread a little too far when he also suggested – following the lead of the ancient Neoplationists Plotinus and Iamblichus, whose works he translated from Greek – making images that could attract and capture planetary powers, a rather questionable thing for an ordained Roman Catholic priest to be doing. Indeed, Ficino can be read as crossing the line at this point from *natural* magic into *spiritual* magic, although he might well have disputed that interpretation. The former used the hidden sympathies in nature, while the latter elicited the help of spiritual beings – the *daimons* and gods of pagan Greek philosophy, or the demons and angels of Christian theology. The former *magia* was unobjectionable, the latter (reasonably enough) drew the condemnation of theologians. Questions were raised about Ficino's orthodoxy, but apparently no actions were taken, since such rituals could be interpreted as entirely physical and medicinal, and thus entirely acceptable. Over a century later, for example, the Dominican friar Tommaso Campanella and Pope Urban VIII used a ritual of lights, colours, smells, and sounds, not unlike Ficino's prescriptions, to counteract any possible ill effects from the temporary loss of healthful solar influences during a solar eclipse that had been predicted to bring about the pontiff's death. The Pope survived. Yet while this *magia* was natural in intended operation, some onlookers did view such applications as suspect.

At the present time, applications of *magia naturalis* and the whole idea of an interconnected world of sympathies and analogies are sometimes dismissed as irrational or superstitious. But this harsh judgement is faulty. It results from a certain smug arrogance and a failure to exercise historical understanding. What our predecessors did was to observe various mysterious and apparently similar phenomena in nature and to extrapolate thence into a more universal statement – a law of nature – about connections and the transmission of influences in the world. This extrapolation led to one tenet that they held that we do not; namely, that similar or analogous objects silently exert influence upon one another. Once that assumption is made, then the rest of the system builds upon it rationally. They were trying to understand the world; they were trying to make sense of things and to make use of the powers of nature. They moved inductively from observed or reported instances to a general principle and then deductively to its consequences and applications. We might choose to say, informed as we are by more recent studies, that the action between Sun and sunflower, or Moon and sea, or magnet and iron, can be better explained by something other than hidden knots of sympathy. But that does not permit us to say that their methods or conclusions were irrational, or that the beliefs and practices that came from them were 'superstitious'. If that leap were allowed, then every scientific theory that comes ultimately to be rejected in the course of the development of our understanding of the world – no doubt including some things that we today believe to be true explanations of phenomena – would have to be judged irrational and superstitious as well, rather than simply *mistaken* notions that were arrived at rationally given the ideas, perspectives, and information available at the time.

Religious motivations for scientific investigation

Magia naturalis is only the strongest expression of widely held ideas of a connected world, of the macrocosm and microcosm, and of the power of similitude. The same kinds of connections and thinking were often implicit in the work of natural philosophers who never gave natural magic a second thought. Every thinker of the period, for example, was confident of the intimate connections among human beings, God, and the natural world, and consequently of the interconnections between theological and scientific truths. This feature brings up the complex topic of science and theology/religion. In order to understand early modern natural philosophy, it is necessary to break free of several common modern assumptions and prejudices. First, virtually everyone in Europe, certainly every scientific thinker mentioned in this book, was a believing and practising Christian. The notion that scientific study, modern or otherwise, requires an atheistic – or what is euphemistically called a 'sceptical' – viewpoint is a 20th-century myth proposed by those who wish science itself to be a religion (usually with themselves as its priestly hierarchy). Second, for early moderns, the doctrines of Christianity were not opinions or personal choices. They had the status of natural or historical facts. Dissension obviously existed between different denominations over the more advanced points of theology or ritual practice, just as scientists today argue over finer points without calling into question the reality of gravity, the existence of atoms, or the validity of the scientific enterprise. Never was theology demoted to the status of 'personal belief'; it constituted, like science today, both a body of agreed-upon facts and a continuing search for truths about existence. As a result, theological tenets were considered part of the data set with which early modern natural philosophers worked. Thus theological ideas played a major part in scientific study and speculation – not as external 'influences', but rather as serious and integral parts of the world the natural philosopher was studying.

Many people today acquiesce in the widespread myth, devised in the late 19th century, of an epic battle between 'scientists' and 'religionists'. Despite the unfortunate fact that some members of both parties perpetuate the myth by their actions today, this 'conflict' model has been rejected by every modern historian of science; it does not portray the historical situation. During the 16th and 17th centuries and during the Middle Ages, there was not a camp of 'scientists' struggling to break free of the repression of 'religionists'; such separate camps simply did not exist as such. Popular tales of repression and conflict are at best oversimplified or exaggerated, and at worst folkloristic fabrications (see Chapter 3 on Galileo). Rather, the investigators of nature were themselves religious people, and many ecclesiastics were themselves investigators of nature. The connection between theological and scientific study rested in part upon the idea of the Two Books. Enunciated by St Augustine and other early Christian writers, the concept states that God reveals Himself to human beings in two different ways – by inspiring the sacred writers to pen the Book of Scripture, and by creating the world, the Book of Nature. The world around us, no less than the Bible, is a divine message intended to be read; the perceptive reader can learn much about the Creator by studying the creation. This idea, deeply ingrained in orthodox Christianity, means that the study of the world can itself be a religious act. Robert Boyle, for example, considered his scientific inquiries to be a type of religious devotion (and thus particularly appropriate to do on Sundays) that heightens the natural philosopher's knowledge and awareness of God through the contemplation of His creation. He described the natural philosopher as a 'priest of nature' whose duty it was to expound and interpret the messages written in the Book of Nature, and to gather together and give voice to all creation's silent praise of its Creator.

In sum, early moderns saw – in various ways – a cosmically interconnected world, where everything, human beings and God and all branches of knowledge, were inextricably linked parts of a whole. In some respects, the recent development of ecology and environmental sciences might be seen as restoring some lines of the unseen networks of interdependence early modern natural philosophers envisioned in their own world. However that may be, early modern thinkers, like their medieval forebears, looked out on a world of connections and a world full of purpose and meaning as well as of mystery, wonder, and promise.

Chapter 3
The superlunar world

Until the modern age, the heavens were quite literally half of people's daily world. The sky and its movements were inescapable. It is ironic and tragic that while modern science now gives us better explanations of the workings of the celestial world than ever before, modern technology means that most people can no longer see its nightly movements with their own eyes, feel its presence, and marvel at its beauty. It now requires an unobscured view far from the pollution of light and industry to witness the impact of the night sky as our ancestors did. Long before the invention of writing, ancient peoples knew the movements of the heavens. Figuring out how to explain these movements, however, occupied acute minds down to the 18th century. The gradual uncovering of the hidden structures of the heavens represents a key narrative of the Scientific Revolution. The best-known names of the era – Copernicus, Kepler, Galileo, Newton – are principal players in this story. Indeed, developments in astronomy stood for a long time as *the* narrative for the period, providing much of the foundation for giving it the title of 'revolution'.

For the intellectual of 1500, the universe was divided into two realms: the *sublunar world* of the Earth and everything up to the Moon, and the *superlunar world* of the Moon and everything beyond. This division had been drawn by Aristotle, based on the common observation of the dichotomy between the unchanging heavens and the ever-changing Earth. In the sublunar world, the four elements of earth, water, air, and fire constantly combine, dissociate, and recombine; new things appear and old things vanish. The superlunar world was quite a different matter; it was the realm of the unchanging. For centuries before Aristotle, stargazers had watched planets and stars follow their courses with perfect regularity. This absence of change suggested to Aristotle that the superlunar world was composed of a single homogeneous substance, a fifth element he called *aither* (later writers called it quintessence), which could neither change nor decompose because it was pure and elemental.

Observational background

The Greeks initiated the long endeavour to *explain* celestial motions physically and mathematically. These motions are more complex and more orderly than most people today recognize. Everyone is familiar with the daily motion of rising and setting. Everything – Sun, Moon, planets, stars – rises and sets once a day, moving east to west across the sky. Other celestial motions demand more patient observation. The stars, called the 'fixed stars' because they do not move relative to one another, take a little less than 24 hours to come back to the same position in the sky. That means that each star rises a little earlier (about four minutes) each night; therefore, if you look at the sky every night at the same time, you will see the constellations moving slowly from night to night in great arcs around – if you are in the Northern Hemisphere – the one star that never moves, Polaris, the pole star, found at the end of the Little Dipper (Ursa Minor). The stars take one year to return to the same place in the sky at the same time of night. The impression is that of a great shell studded with stars, turning around the Earth once every 23 hours and 56 minutes.

The Sun moves a little more slowly, taking a full 24 hours for one revolution, meaning that from day to day it changes its position relative to the stars, moving slowly *west to east relative to the backdrop of the stars*, taking one year before it lines up with the same stars again. The Moon makes a similar motion, but much more noticeably. It rises about 50 minutes *later* each night, so if you look for it at the same time on consecutive nights, you will find it further to the east every night. (Go ahead and try it!). After 29 days, the Moon is back where it started. The planets do the same thing, but with a weird twist that screams out for explanation. Most of the time they act like Sun and Moon, moving slowly west to east against the backdrop of stars. But at intervals, they slow down, stop, turn around, and move in the opposite direction, going now east to west. This is called *retrograde motion*. After a while, they stop again, turn around, and resume their usual motion.

The ancient Greeks gave the name 'planet' (meaning 'wanderer') to all seven heavenly bodies that appeared to move against the fixed background of stars: the Sun, the Moon, Mercury, Venus, Mars, Jupiter, and Saturn. But the planets don't wander far; their movements are restricted to a narrow band in the heavens called the zodiac. The zodiac is divided into twelve sections of equal length, each containing a single constellation or 'sign': Aries, Taurus, Gemini, and so on. Thus, as the planets make their individual motions against the backdrop of the stars, they appear to move through the zodiac from one constellation and from one sign to the next. A person's 'sign' is whatever zodiacal sign the Sun was 'in' on the day the person was born. But more on astrology in a little while.

Historical background

Plato was convinced that the heavens moved according to harmonious mathematical laws. He was inspired by the ideas of the Pythagoreans, a secretive religious community, who taught that mathematics – number, geometrical shape, ratio, and harmony – was the proper foundation of both the universe and the well-governed life. For Plato and those he inspired down to the modern age, the Creator is a geometer. But the irregular motions of the planets seemed discordant with the idea of a well-regulated mathematical world. Plato therefore argued that their motion only *appears* irregular, and that there exists a divine regularity hidden from our eyes. Because he considered the circle to be the most perfect and regular shape, and motion in a circle to be without beginning or end and thus eternal, he challenged his students to explain the apparent motions of the planets using combinations of *uniform circular motions*. That challenge inspired astronomers for over two thousand years.

Plato's student Eudoxus proposed a universe built up of concentric spheres, like layers of an onion, with the Earth at the centre. Each sphere rotated uniformly, but each planet received the combined motion of several spheres, that added up (approximately) to the observed motion. Eudoxus' system was a *mathematical* model. He did not worry about how the heavens worked physically, or whether there really were spheres up there. The point was to account for observations mathematically. Aristotle, however, wanted a *physical* model. He made Eudoxus' spheres real, solid objects that literally carried the planets around, and accounted for how motion could be transferred from one sphere to the next, like gears of a celestial clockwork. His achievement was to construct an astronomy and physics that worked together harmoniously (Figure 2).

The problem with the concentric spheres model was that it failed to explain observations accurately. For example, the planets change in brightness, as if they were sometimes closer and sometimes farther away, and the seasons are not of equal length. Neither is it possible if the planets are carried by spheres centred on the Earth (Figure 3).

Later astronomers addressed these problems, culminating in the system of Claudius Ptolemy (c. AD 90–c. 168). To solve the problem of unequal seasons, Ptolemy used an *eccentric*; that is,

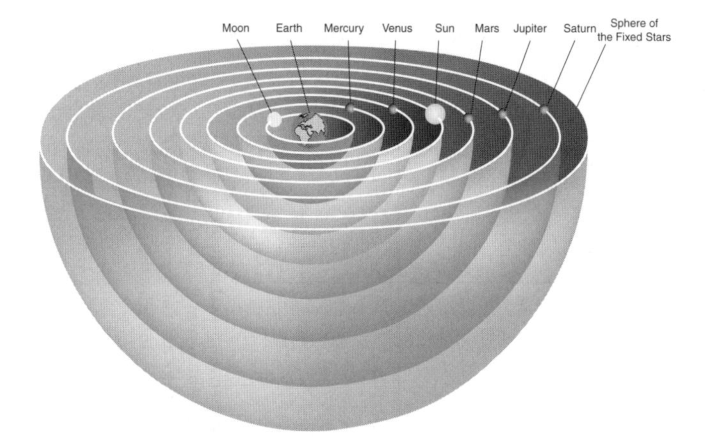

2. A simplified version of Aristotle's concentric spheres model in cross-section

he moved the Earth off-centre. In his system, each sphere has its own centre, none of them coincident with the Earth.

To account for planetary positions better and to solve the problem of their changing brightness, Ptolemy used *epicycles* (Figure 4). Each planet moves in a small circular path centred on, and carried around by, a larger sphere (the deferent) around the Earth. The motions of epicycle and deferent combine to give the planet a looping path that explains observed motions extremely well, and in which the planet is sometimes closer to the Earth, hence brighter.

Ptolemy's system gave good predictions of planetary positions but satisfied the mathematically inclined more than the physically inclined. Aristotle's physics held that heavy bodies fall towards the centre of the universe, which is why a spherical Earth occupies that space and why heavy objects fall. But Ptolemy's Earth is

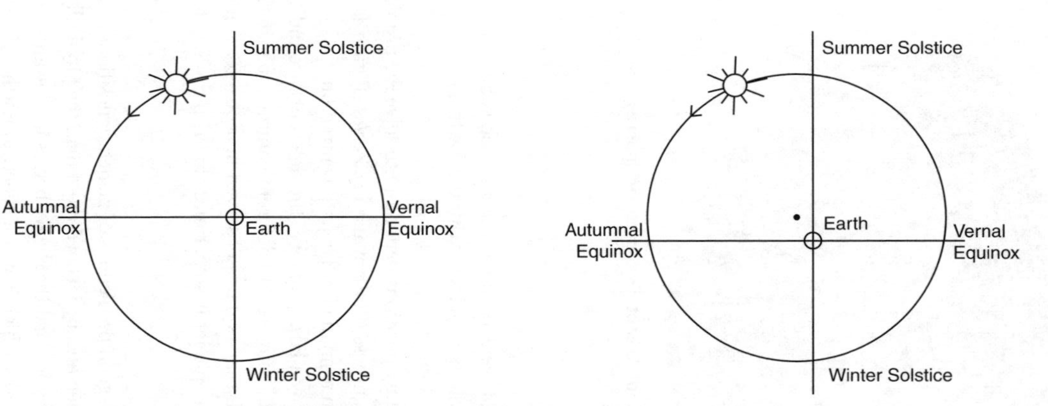

3. (left) If the Earth were at the centre of the Sun's sphere, the Sun's apparent annual motion would divide into four equal arcs, making the seasons have equal lengths. But summer is in fact longer than winter; (right) Ptolemy's off-centre Earth divides the Sun's path into four arcs of unequal length, corresponding correctly with the unequal seasons. This arrangement also explains why the Sun appears to move more slowly in the summer: because it is farther away then

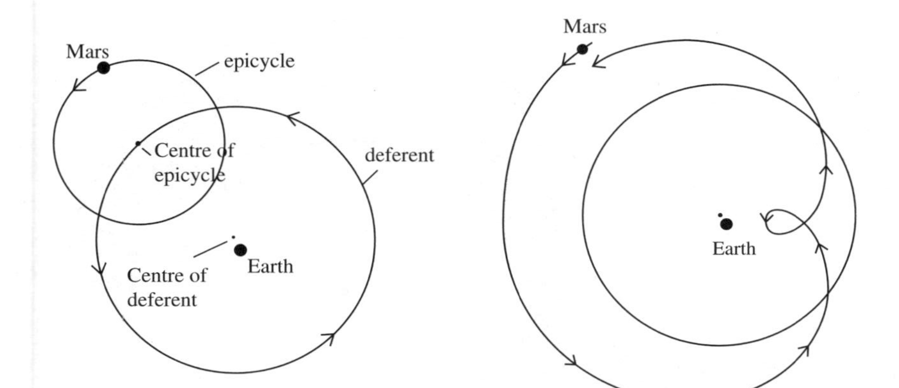

4. (left) A Ptolemaic epicycle and deferent for a planet. The planet moves counterclockwise (looking down at the Earth's north pole) on the epicycle, as the epicycle is carried around on the deferent, also counterclockwise; (right) The apparent motion of the planet resulting from the combined motions of epicycle and deferent. When the planet is outside the deferent, it appears dimmer and moves west to east; inside, it appears brighter because it is closer and at closest approach it moves east to west (retrograde)

THEORICAE NOVAE PLANETARVM GEORGII PVRBACHII ASTRONOMI CELEBRATISS.

DE SOLE

Ol habet tres orbes a se iuicẽ omniquaq; diuisos atq; sibi cõtiguos Quoꝝ supræ/mus secũdũ superficiẽ conuexá est mũdo cõcentricus: secũdũ cõcauá aũt eccẽtricus Infimns uero secũdũ cõcauá cõcentric⁹: sed secũdũ conuexá eccẽtric⁹ Tertius aũt i hoꝝ medio locatus tam secũdũ super/ficiem suá conuexá q̃; concauá est mũdo eccentric⁹. Dicit' aũt mũdo cõcẽtric⁹ or/

THEORICA ORBIVM SOLIS.

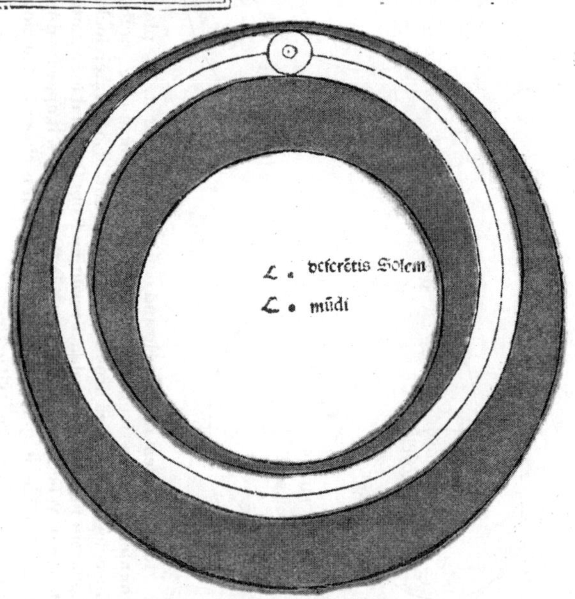

5. An adaptation of Ibn al-Haytham's thick-spheres model popularized by Georg Peurbach and included in 15th-century and later editions of the standard textbook of astronomy, Sacrobosco's *The Sphere* – this image from the 1488 Venice edition shows the Sun's sphere

off-centre; why doesn't it move to the centre? Why would heavy objects fall to something other than the centre? This discrepancy between the mathematical model and the physical system vexed medieval Arabic authors while both Aristotle's and Ptolemy's works were unknown in Europe. Ibn al-Haytham, or al-Hazen (c. 965–1040) adopted a compromise. His system had spheres centred on the Earth, which kept physicists happy. But these spheres were thick and solid enough to contain circular tunnels not centred on the Earth, through which the planets moved on their epicycles and deferents, which accounted for observations (Figure 5).

Medieval European astronomers inherited these ideas and problems and, like their Arabic colleagues, continued to refine and update the system, striving to maintain the best accuracy in predicting planetary positions or, somewhat less frequently, trying to generate a physically satisfactory system.

Early modern astronomical models

Nicholas Copernicus (1473–1543) spent most of his life as canon – an administrative post in Holy Orders – for the cathedral church in Frauenburg (today Frombork, Poland). He studied canon law in Bologna and medicine in Padua, and earned a doctorate in law at Ferrara in 1503. While at Bologna he began studying astronomy, and around 1514 he wrote an outline of his idea that the Sun, not the Earth, was at the centre of the planetary system. In his *heliocentric* (Sun-centred) system, the Earth rotates on its axis once a day, producing the familiar appearance of the entire cosmos turning around the Earth. What seems to be the motion of the Sun through the zodiac is really an illusion caused by the Earth's motion around the Sun. The observed 'loop' and retrograde motion of Mars, Jupiter, and Saturn does not result from their own motion, but rather from a combination of *ours and theirs* whenever the Earth laps one of these planets in the race around the Sun (Figure 6). Only the Moon revolves around the Earth.

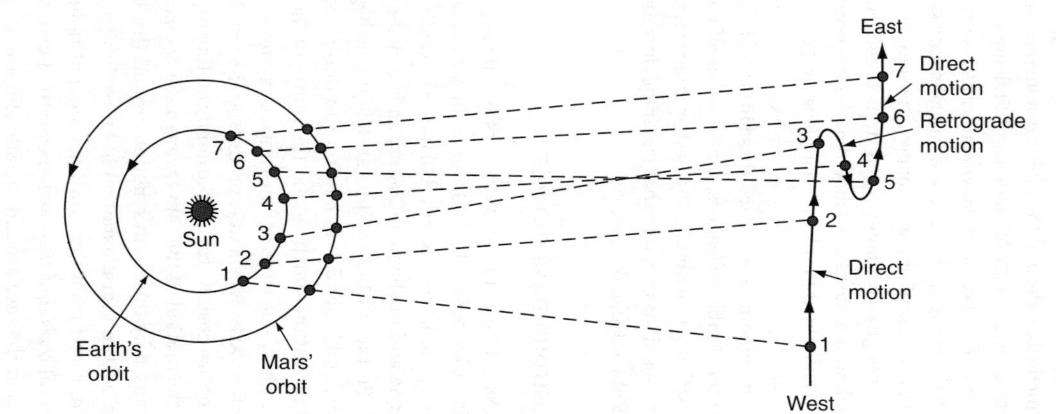

6. Copernicus's explanation of retrograde motion for one of the 'superior', or outer, planets (Mars, Jupiter, or Saturn). The 'loop' is an illusion caused when the Earth moves past one of these planets

Copernicus's work circulated in manuscript, and sufficiently established his reputation as an astronomer that in 1515, when a Church council was considering how to reform the old Julian calendar - in use since the Romans and now in need of an overhaul - they wrote to ask Copernicus's opinion. (Copernicus told them that the length of the solar year needed to be established more accurately first.) But Copernicus was reticent to publish a complete exposition of his system. He kept refining it for over 25 years, and had it not been for the nagging of several prominent churchmen it might never have been published. In 1533, for example, Johann Albrecht Widmannstetter, the Pope's personal secretary, lectured on Copernicus's system to the delight of Pope Clement VII and several cardinals. The cardinal of Capua, Nicolaus Schönberg, wrote to Copernicus saying that:

> I have learned that you teach that the Earth moves; that the Sun occupies the lowest, and thus the central place in the world... and that you have prepared expositions of this whole system of astronomy.... Therefore I most strongly beg that you communicate your discovery to scholars.

But Copernicus continued to demur, kept busy with his duties as canon, and expressing fear of criticism over his system's novelty.

In 1538, a young professor of astronomy named Georg Joachim Rheticus, sent from the University of Wittenberg by Melanchthon, came to study with Copernicus. Rheticus compiled and published a summary of Copernicus's ideas; the response was sufficiently positive that Copernicus finally agreed to publish his full manuscript, and gave it to Rheticus to shepherd through the press. Rheticus embarked on the task, but then took a job in Leipzig and handed off the project to a Lutheran minister named Andreas Osiander. Osiander finished the publication, and *On the Revolutions of the Heavenly Orbs* finally appeared in 1543 - a copy reaching Copernicus just before he died.

The book's appearance did not unleash the criticism Copernicus feared. It was read, but few readers were convinced. There were probably no more than about a dozen convinced Copernicans for the rest of the century. Why? Copernicus's heliocentric system did not fit observational data any better than the geocentric, nor was it physically much simpler. In fact, Copernicus had to keep using epicycles and an off-centre Sun in order to make his system harmonize with observations. More seriously, a moving Earth conflicted with basic physics, common sense, and possibly Scripture. Heavy bodies, like the Earth, naturally fall to the centre of the universe, its lowest point – this principle of 'natural place' explains why heavy objects fall. So how could the entire Earth remain suspended so far from the centre? Common sense indicates we are not moving. To rotate once a day, the Earth would have to turn very fast, yet we have no sensation of motion and neither birds in flight nor clouds are left behind by an Earth spinning beneath them. Some medieval thinkers had discussed the possibility of a rotating Earth. Nicole Oresme (c. 1325–82) concluded that all motions are relative, and without a point of reference it is impossible to decide whether the Earth is spinning or the heavens revolving. But it seems more likely, he concluded, that the Earth is stable and the heavens moving. More literal readers of Scripture could cite passages that speak of a stable Earth and a moving Sun, although interpretations varied widely. Finally, if the Earth moves around the Sun, the stars should exhibit parallax – a small shift in their apparent relative positions as the Earth swings from one side of its orbit to the other. But no parallax could be detected, meaning that either the Earth was *not* moving, or the stars were *incomprehensibly* far away. The 13th-century Campanus of Novara estimated that Saturn's sphere was about 73 million miles away – a staggering distance to even the best-travelled medievals – and the fixed stars lay just beyond. Copernicus estimated Saturn's sphere to be about 40 million miles away, but the lack of stellar parallax meant (according to later calculations) that the stars would have to be at least 150 *billion* miles farther out. This enormous gulf of emptiness seemed absurd to Copernicus's readers. (In fact, the closest star is 170 times farther away than the most modest prediction made from the lack of visible parallax. Stellar parallax was not detected until 1838.)

Several factors seem to have convinced Copernicus of heliocentrism, even without observational evidence. In his dedicatory letter to Pope Paul III, Copernicus referred to the Ptolemaic system, with its eccentrics, epicycles, and treatment of each planet separately, as a 'monster'. Noting that the world is 'created by the best and most systematic Artisan of all', it should be harmonious. Copernicus, as the humanist he was, saw himself as clearing away later 'accretions' to return to Plato's original challenge of showing the well-ordered nature of celestial motions. Worried about the 'novelty' of his system, he tried to minimize the appearance of novelty by citing ancient precedents – Aristarchus of Samos, Pythagoras, a certain Nicetas mentioned by Cicero – and even reinterpreting some Bible passages to favour heliocentrism.

One could, however, appreciate Copernicus's system without believing it to be true. Tables for determining planetary positions were easier to calculate in a heliocentric system; therefore, some astronomers adopted it as a 'convenient fiction'. Copernicus himself presented heliocentrism as a true description of the world, but Osiander undercut him by surreptitiously adding his own (unsigned) preface to Copernicus's book. Osiander wrote that we are 'absolutely ignorant of the true causes of planetary motions' and that:

> it is not necessary that these hypotheses be true or even probable; one thing suffices, that they give calculations matching observations... let no one expect anything certain from astronomy since it can give no such thing, nor should he take up anything confected for another purpose as if it were truth, lest he leave this study more stupid than he arrived.

Had Copernicus not already suffered a stroke, he might have had one when he saw Osiander's words. Rheticus was furious, and in his copy of the book scratched out Osiander's preface. The tension is once again between mathematical models and physical systems. Most astronomers were interested primarily in where planets would be when; whether the Sun went around the Earth or the Earth around the Sun simply did not matter, and many doubted whether one could ever tell for sure which was true. It was enough for an astronomical theory to provide tables and calculations to get planetary positions right. For the majority, practical results trumped theory. To understand this situation, we have to realize that the major driving force behind astronomical studies since before the time of Ptolemy was astrology, a practical enterprise that required being able to calculate planetary positions down to the minute, often many years in the past or the future.

Practical astronomy, or, astrology

Astronomy ('laws of the stars') measured and calculated the positions of heavenly bodies and hypothesized cosmological systems; astrology ('study of the stars', compare with geology, biology, and so on) endeavoured to explain and predict the heavenly bodies' effects on Earth. In general, these two endeavours – the first theoretical, the second practical – were pursued by the same people. Many early modern astronomers made their living primarily from practising astrology. Do not confuse ancient, medieval, or early modern astrology with the inanities of 'newspaper horoscopes'. Astrology was a serious and sophisticated practice based on the idea that heavenly bodies extend certain influences to earth – a key part of the conception of an interconnected world. Most medieval and early modern astrology is not 'magical', supernatural, or irrational; it depends upon natural mechanisms that are simply part of the way the world is put together. Light reaches us from the planets, so why shouldn't some additional influence accompany that light, just as the light from a fire also warms objects at a distance? Celestial influences on Earth are easy to observe – the Moon's link with the tides, or the Sun's zodiacal position with seasonal weather. Effects on the human body are clear as well, such as the synchronicity of the lunar cycle with menstruation. The reality of celestial influences seemed too obvious to question; the many controversies over astrology involved rather the extent of these influences and how to predict their effects accurately. The system of crisscrossing influences from seven planets constantly changing their positions relative to one another ('aspects'), and forever moving through twelve zodiacal signs which were themselves unceasingly passing through twelve 'houses' (positions relative to the horizon), made for an incredibly complex system. The complexity of indications and counterindications, knowns and unknowns, is comparable to the modern task of identifying factors in global climate change or predicting future economic trends. Relative to the latter enterprise, early modern astrologers possibly had a better success rate.

Astrology included several overlapping branches. Meteorological astrology endeavoured to predict the weather for the coming year. Many practitioners – often simply called 'mathematicians', a testimony to the calculations required for astrology – made a living producing almanacs containing calendars, lunar cycles, dates of eclipses, weather predictions (like the *Farmers' Almanac* today), and prognostications of important events or trends. The printing press made these publications inexpensive and widely distributed. Physicians used medical astrology to suggest crucial times for treatments and the course and possible causes of illnesses (see Chapter 5). Natal astrology used planetary positions at the exact time and place of a person's birth in order to determine what influences they 'imprinted' on the newborn. The specific combination of planetary influences would produce a unique 'complexion', or innate constitution, in the humoral system, leading to particular tendencies and traits. These tendencies (proneness to certain illnesses, to anger, laziness, melancholy, and so on) could be temporarily enhanced by subsequent planetary alignments. The goal of such astrology was thus to obtain information about a person's natural constitution, in order to be aware of particular strengths and weaknesses, and to provide advance notice of potentially dangerous or salubrious times. In stronger forms, this practice shaded into a judicial astrology that was criticized as unacceptably deterministic, namely, that astrological influences direct our actions and fates. Theologians condemned such notions as a violation of human free will. The scholarly consensus in the early modern period was that 'the stars incline but do not compel' us, and that *sapiens dominatur astris* ('the wise man rules the stars'). In short, human beings can always choose their actions, although the completely free exercise of will could be subject to external influences (such as a diminished capacity for reason owing to temporary irascibility due to a humoral imbalance caused by a particular position of Mars). Indeed, a parallel can be drawn between early modern astrology and current 'nature versus nurture' debates in their mutual attempts to explain human behaviour. The notable difference, ironically, is that moderns seem to have forgotten about the primacy of free will.

Some judicial astrology attempted to identify propitious dates for important endeavours. The mathematician and magus John Dee (1527–1608/9) used astrology to choose the best coronation day for Elizabeth I. A horoscope was cast for the founding of the Lincei, one of the first scientific societies, and also for the date of setting the cornerstone of the new St Peter's in Rome. Some astrological dates were chosen not to take advantage of a favourable 'influence', but rather to add levels of meaning to an event, the way for example, American scientists chose the landing date for probe on Mars to coincide with US Independence Day. Other forms of judicial astrology endeavoured to predict future events – such as wars and deaths – thus potentially moving away from the *natural* causality whereby learned early modern astrology was considered to operate. One way around this problem was to consider certain celestial events, comets in particular, not as *causes* but rather as *portents*, divinely sent signs of things to come. The interest in celestial portents was more pronounced in northern, that is, Protestant Europe, partly thanks to a preface written by Philipp Melanchthon for Protestant editions of Sacrobosco's *Sphere* – a fundamental astronomy textbook – in which he underscored the importance of astrology for reading God's signs in the heavens. In sum, astrology of various sorts was a source of helpful information for better living; its ubiquity in early modern thought emphasizes how the superlunar world was truly half of people's daily world.

Heavenly changes and divine harmonies

Astrological concerns over a heavenly portent led to the debut of the Danish astronomer and nobleman Tycho Brahe (1546–1601). In November 1572, he saw a bright object in the constellation Cassiopeia where none should be. Tycho was astonished – what could this object be, and what did it mean? In his astrological almanac for 1573, Tycho struggled to explain the object, concluding it was a divine portent of tumultuous changes to come. Tycho watched this brilliant point of light, but it did not move like a comet would. He and others around Europe tried to measure its diurnal parallax in order to determine its distance, but they found none, meaning that it was much farther away than the Moon – in the superlunar world, the realm thought free from change, yet it was a *new* star. (What Tycho saw was a supernova; the expanding remnants of that cataclysmic detonation were located in 1952. The term *nova* comes from Tycho's Latin for this object – *stella nova*, or new star.)

Soon after, in 1577, a bright comet appeared. Aristotle had taught that comets, like meteors, were sublunar phenomena caused by the ignition of flammable exhalations in the upper atmosphere. As erratic, changing objects, they could have no place in the unchanging superlunar world. Astrologically, Tycho concluded that the comet of 1577 continued the warning given by the new star, but this time he detected a diurnal parallax. His measurement, confirmed by others, indicated that the comet was far beyond the Moon, in the sphere of Venus. Tycho observed the same thing in 1585 when another bright comet appeared. These comets provided further evidence of change in the 'immutable' heavens, and their positions indicated that they were passing *through* the planetary spheres, implying that there were no solid spheres moving the planets. What then kept the planets in their regular courses? This puzzling liberation of the planets from solid spheres meant that the paths of celestial objects could cross one another, which in turn liberated Tycho to devise a new system of the heavens that combined his observations with the best parts of Ptolemy and Copernicus, while avoiding objectionable parts of both. In Tycho's *geoheliocentric* system, the Earth remained at rest at the centre, as common sense and Scripture dictated, with the Moon in orbit around it. The planets however all circled the Sun, which moved with its planetary retinue around the Earth.

While Tycho, within the castle-observatory Uraniborg he built on the island of Hven in the Danish sound, continued to observe and make the most accurate measurements of the heavens ever performed, Johannes Kepler (1571–1630), a convinced Copernican, was making his own astonishing discoveries on paper. In the 1590s, while teaching at a high school in Graz, Kepler puzzled over a question modern scientists would not think to ask. In Copernicus's system, there were only six planets orbiting the Sun, no longer seven orbiting the Earth. Seven planets had matched up nicely with the seven days of the week, the seven known metals, the seven tones of the musical scale, and all the other significant sevens in the world. Seven planets had a pleasing harmony, appropriate for an interconnected world; six did not. Why then were there six and only six planets, and why had God placed them at the distances He did? In the early modern world, a world full of meaning and purpose, everything has a message to be read.

While lecturing on 19 July 1595, Kepler suddenly realized that if one inscribes a regular polygon (triangle, square, pentagon, etc.) within a circle, and then inscribes a circle within that polygon, one obtains two circles whose relative sizes are fixed by the choice of polygon. Excitedly, he began calculating the ratios determined by different polygons to see if any of them matched up with the ratios of planetary distances from the Sun. They did not. Undaunted, he tried spheres and polyhedra instead of circles and polygons. In this case, by nesting spheres and polyhedra in the right order, Kepler obtained spheres whose sizes matched the distances of planets from the Sun estimated by Copernican theory. Moreover, since there exist only five regular polyhedra (that is, solid bodies where all the faces are identical, the five so-called Platonic solids: tetrahedron, cube, octahedron, dodecahedron, and eicosahedron) to use as spacers, there can be *six and only six* spheres, and thence, exactly *six* planets. For Kepler, this was an awesome discovery. He had found the cause for the number and distances of planets, and uncovered a geometrical structure to the heavens whose elegant beauty served as the best proof yet of the Copernican system. This striking correlation could not be random; Kepler had uncovered the mathematical blueprint by which God constructed the heavens.

Kepler exemplifies the unity of human inquiry normal for the early modern period. Theological and scientific inquiry are not separate: to study the physical world means to study God's creation, to study God means to learn about the world. Indeed, Kepler became convinced of Copernicanism partly because the heliocentric universe provided a physical analogy to the Holy Trinity: God the Father symbolized by the central Sun, God the Son by the sphere of the fixed stars that receives and reflects the Sun's rays, and God the Holy Spirit, theologically the love between Father and Son, by the light-filled space between the two. Drawing upon the idea of the Two Books, Kepler and his contemporaries were certain that God built messages into the fabric of creation for man to uncover. Thus theological motivations – the desire to read those messages in the Book of Nature – provided the single greatest driving force for scientific inquiry throughout the entire early modern period.

Kepler announced his discovery in the *Cosmographic Mystery* (1596) and sent a copy to Tycho Brahe. Tycho invited Kepler to collaborate; Kepler initially declined, but after Tycho moved to the court of Emperor Rudolf II in Prague as Imperial Counselor, Kepler joined him there in 1600. When the Danish nobleman died the next year, the Emperor made Kepler his Imperial Mathematician. Tycho had set Kepler studying the motions of Mars, and after a long struggle endeavouring to give it a path that fit the positions Tycho observed, Kepler came to a startling conclusion. He found that the planet's positions could be accounted for best by making it move on an *ellipse* not a circle. Kepler thus, reluctantly, broke with two thousand years of astronomical tradition focused on circles. But since (in Kepler's words) Tycho had 'smashed the crystalline spheres', what kept the planets moving in elliptical paths? Kepler postulated a 'moving soul' (*anima motrix*) in the Sun, a power that pushed the planets along. Like the Sun's light, this force decreased with distance, hence the further a planet was from the Sun, the more slowly it was moved. Drawing on the recent claim of William Gilbert (1544–1603) that the Earth is a gigantic magnet (see Chapter 4), Kepler postulated a second solar force, analogous to magnetism, that attracted the planets at some points and repelled them at others. The combination of *anima motrix* and 'magnetic' virtue kept the planets moving in ellipses, without the need for governing spheres, moving faster when they were pulled closer to the Sun and more slowly when they were pushed further away. Even as Kepler abandoned uniform circular motion, he was delighted to detect another uniformity to replace it: the 'equal area law', namely, that a line from the Sun to a planet sweeps out equal areas in equal times as the planet moves. Likewise, even as he helped dismantle Aristotle's cosmos, Kepler subtitled his *Epitome of Copernican Astronomy* as a 'supplement' to Aristotle's *On the Heavens*. Both continuity and change, both innovation and tradition, characterize early modern natural philosophy.

Telescopes and the Earth's motion

Tycho had been the greatest naked-eye observer; he was also among the last. While Kepler calculated, Galileo Galilei (1564–1642) heard of a Dutch device that made distant objects appear closer, built an improved one for himself, and turned it to the heavens in 1609. Virtually everywhere he directed his *occhiale* (later called the telescope) he made new discoveries. He found the Moon's surface covered with mountains, valleys, and oceans – in other words, looking just like the Earth and therefore made of the same four elements, not the Aristotelian quintessence. He found four moons orbiting Jupiter, like a little planetary system, and earned himself a patron and promotion by naming them the Medicean stars after Cosimo II de' Medici, Grand Duke of Tuscany. He found that Saturn had a strange shape that looked to him like three spheres joined together. He found that the planet Venus showed phases like the Moon. This last discovery was the first solid evidence against the Ptolemaic system, in which Venus could never be more than a crescent since it would always lie between the Sun and the Earth. Galileo's observation of both a crescent *and* a full Venus proved that it must sometimes be between us and the Sun and sometimes on the far side of the Sun, in short, that it orbits the Sun. Henceforth, astronomers would have to choose between versions of the Tychonic or the Copernican systems (Figure 7). The question of the mobility of the Earth – the only point that divided Tycho from Copernicus – thus came to be of prime concern.

Galileo rushed his first telescopic discoveries into print as the *Starry Messenger* (1610), sending them out to astronomers and rulers throughout Europe along with telescopes. Many had difficulty seeing what he described because the magnifications were low, the optics poor, and the telescope difficult to use. A key endorsement came from the Jesuit astronomers in Rome, who confirmed and continued Galileo's observations, and gave a feast

7. Three world systems compared in the emblematic frontispiece of Giovanni Battista Riccioli's *Almagestum novum* (Bologna, 1651). Astraea, goddess of justice, weighs the systems of Copernicus and Riccioli (a slight modification of Tycho's) while Ptolemy reclines alongside his now-discarded system. Above, cherubs carry the planets showing recent discoveries: the phases of Mercury and Venus, the Moon's rough surface, Jupiter's satellites, and Saturn's 'handles'. The divine hand blesses the world, its three extended fingers marked 'number, weight, measure' (Wisdom 11:20) expressing the mathematical order of creation

in his honor in 1611. The senior member of the Collegio Romano, Christoph Clavius (1538–1612), one of the most respected mathematicians in Europe and the man who had devised the new Gregorian calendar put into place in 1582 by Pope Gregory XIII (and still in use today), wrote that Galileo's discoveries required a rethinking of the structure of the heavens. Although Clavius and many others maintained geocentrism, some younger Jesuit astronomers were probably converted to heliocentrism. These excellent relations, however, would not weather Galileo's disputes (in which he often became insulting) with two Jesuit astronomers: Christoph Scheiner over the priority of discovery and nature of sunspots, and Orazio Grassi on comets (Grassi supported Tycho's measurements that comets were celestial bodies, while Galileo insisted they were sublunar optical illusions).

There is no episode in the history of science more subject to mythologizing and misunderstanding than 'Galileo and the Church'. The events resulted from a tangle of intellectual, political, and personal issues so intricate that historians are still unravelling them. It was *not* a simple matter of 'science versus religion'. Galileo had supporters and opponents both inside and outside Church hierarchy. The events are tied up with offended feelings, political intrigue, who was qualified to interpret Scripture, being at the wrong place at the wrong time, and being caught between Church factions. The final trigger was Galileo's 1632 publication of the *Dialogue on the Two Chief World Systems*, which compared Ptolemaic and Copernican systems, obviously choosing the latter as true, and the Earth as mobile. Galileo's chief evidence was his notion that the Earth's motion caused the tides; in this, he was famously wrong despite being right about the Earth's motion. The Church had no direct stake in which system was true; neither geocentrism nor Aristotelianism was ever Church dogma. But the Church did have a stake in biblical interpretation, and not only did a moving Earth have implications for interpretation, but Galileo had rather rashly dabbled in that matter in the early 1610s to support his ideas. This looseness with Scripture resembled the licence being taken contemporaneously by Protestants to reject traditional interpretations in favour of their own personal ones. As a result, in 1616 Galileo was told, and agreed, to treat heliocentrism and the Earth's motion hypothetically and not as literally true until there was demonstrable evidence. In 1624, Galileo got from his friend Maffeo Barberini, now Pope Urban VIII, permission to write the *Dialogue*, provided that Galileo include the Pope's methodological argument that a natural phenomenon (like the tides) might have several possible causes, some of which may be unknowable, and so we cannot assign it a single cause with absolute certainty. Galileo complied, but then put the argument only on the last page of the book, in the mouth of the character made to play the fool throughout. Galileo also 'neglected' to tell Urban about his 1616 agreement. When the book appeared (with the approval of the Vatican's licensers and censors) and these facts came to light, Urban became furious, feeling that he had been deceived and humiliated. To make matters worse, this petty annoyance materialized while Urban was overwhelmed by diplomatic negotiations regarding the on-going Thirty Years War, mounting criticism, attempts to depose him, and rumours of his impending death. The Inquisition worked out a plea bargain for Galileo that would have sent him home with a slap on the wrist, but the angry Pope refused to accept it – he insisted on making an example of Galileo. Galileo was ordered to abjure the Earth's motion, which he did, and his book was suppressed. Significantly, several cardinals, including Urban's nephew, refused to sign the sentence against Galileo. Galileo was never – folklore aside – condemned as a heretic, imprisoned, or chained.

Galileo ended up under house arrest at his villa in the Tuscan hills. There he continued to work and train students, and wrote perhaps his most important book, the *Two New Sciences*. The impact of his sentencing is difficult to assess. On the one hand, it made some natural philosophers reticent to express Copernican convictions. News of Galileo's condemnation caused René

Descartes (1596–1650), for example, to suppress a recently completed book that embraced heliocentrism. Those in Catholic Holy Orders, like the Jesuits, were now unable to support Copernicanism openly, and so embraced the Tychonic system or variations upon it (Figure 7), although sometimes with a wink and a grin. On the other hand, scientific inquiry, including in astronomy, continued in Italy and other Catholic countries, although sometimes skirting sensitive topics.

Following the conceptual upheavals of two previous generations, the mid-17th century witnessed more observational and technical developments in astronomy than theoretical ones. The French priest Pierre Gassendi (1592–1655) became the first person to witness a transit of Mercury across the Sun's disc in 1631; the event had been predicted by Kepler, who had died in 1630. Improved telescopes led to new discoveries and better measurements, but the need to avoid the distortions from spherical and chromatic aberration meant that telescopes had to be made longer and more unwieldy, sometimes over sixty feet in length. Nevertheless, the odd shape of Saturn was resolved into a ring system, and its largest moon discovered by Christiaan Huygens (1629–1695) in 1656. Gian Domenico Cassini (1625–1712), working in Paris and aided by the superior telescopes made by the Roman optician Giuseppe Campani, added four more, and named them Ludovican Stars after Louis XIV. The Jesuit Giovanni Battista Riccioli (1598–1671) produced a new star catalogue, and with his confrère Francesco Maria Grimaldi (1618–1663), a detailed lunar map providing many of the names still used today for its features – including naming one of the most prominent craters after Copernicus. In Gdansk, Johann Hevelius (1611–87), probably the last person to make careful measurements with both the telescope and the naked eye, also prepared a lunar map, as well as observing comets and participating in the Europe-wide discussion of whether their motion was rectilinear or curved into an orbit around the Sun.

The problem of how the planets keep moving in constant orbits without the aid of solid spheres continued to attract speculation. Descartes proposed a comprehensive world system that became one of the most important of the 17th century. He envisioned all space to be filled with invisibly small particles of matter. These particles moved always in circular streams or vortices. Our solar system was a gigantic vortex of these particles that carried the planets along like a whirlpool carries along bits of straw. This vortex model neatly explained why the planets all move in the same direction and nearly in the same plane. The Earth itself lay at the centre of a smaller vortex that kept the Moon moving in its orbit, and the swirling of matter around the Earth provided a 'wind' that pushed objects toward the centre of the Earth, thus producing the phenomenon of gravity. Descartes' vortex theory gave a comprehensible explanation for celestial motions, and it was widely disseminated in popular treatments and textbooks, but it remained too imprecise to be of practical value for astronomers.

A young Isaac Newton (1643–1727) embraced the Cartesian vortex theory. As a student at Cambridge in the early 1660s, Newton studied the Aristotelian works that remained standard undergraduate texts in most universities. But he soon began an extracurricular reading of the ideas of the 'moderns' such as Descartes. He adopted a modified version of Descartes' principles for both planetary motions and gravity. But by the early 1680s, Newton had begun to think differently. He discarded Cartesian vortices and began thinking in terms of an attractive force existing between Sun and planets. He had at his disposal several sources for this idea, most notably the familiar phenomena of magnetism and the 'magnetism-like' force between Sun and planets that Kepler postulated. For Kepler, the combination of the *anima motrix* and this 'magnetism' produced the planets' elliptical orbits. For Newton, it would be the balance between inertia (the tendency of the planet to move in a straight line tangent to its orbit) and the force of attraction (what we call gravitation) towards the Sun that produced stable elliptical orbits. Several members of the Royal

Society of London had been working along similar lines to explain planetary motion, most notably Robert Hooke (1635–1703), who wrote about his ideas to Newton in 1679–80. Hooke's subsequent complaint that Newton had taken his idea without giving him sufficient credit led the neurotically hypersensitive Newton to expunge any reference to Hooke from his writings and to treat him antagonistically for the rest of his life. Newton's great achievement, published in the *Mathematical Principles of Natural Philosophy* (1687), was to rederive purely mathematically the laws of planetary motion that Kepler had derived empirically from Tycho's observations, and to make gravitation truly universal – that is, existing mutually between all parcels of matter. Kepler would undoubtedly have been pleased; here was more evidence of the harmonious mathematical plan upon which God had created the world. Newton's law of universal gravitation obliterated the last traces of the former distinction between terrestrial and celestial physics – the same law governed the revolution of the planets and the fall of an apple.

Not everyone was pleased. By reviving the idea of attractive forces, Newton seemed to be resuscitating an idea unpopular for some 70 years. An invisible, immaterial force without mechanism or identifiable cause that operated between all bodies was not only less comprehensible than material Cartesian vortices, but seemed to many as a return to the 'hidden qualities' of Aristotelians or the sympathies of natural magic. Indeed, the cutting edge of natural philosophy in the second half of the 17th century had been endeavouring to explain what appeared to be attractions and sympathies by means of a mechanical exchange of invisible particles (see Chapter 5); now Newton seemed to be turning back the clock.

Gottfried Wilhelm Leibniz (1646–1716), with whom Newton waged a priority dispute over the invention of calculus, accused Newton's 'hidden attractive quality' of 'confounding the principles of true philosophy' and returning it 'to the old asylums of ignorance'. While Newton's apologists asserted that gravitational attraction was simply a fundamental property of matter, Newton himself did want to find gravity's *cause*. His method of pursuing that answer, however, reminds us that Newton was not some 'modern scientist' accidentally born in the 17th century. Newton, perhaps with uncharacteristic modesty, considered himself to be only the rediscoverer of the law of universal gravitation; it had been known to the ancients. For Newton believed in the *prisca sapientia*, an idea popular among many Renaissance humanists of an 'original wisdom' divinely revealed aeons ago and corrupted over time. He strove to interpret Greek myths, biblical passages, and the *Hermetica* to show that they concealed ideas about the hidden structure of the world, including his own inverse-square law of gravity. Newton seems to have thought – and believed the 'ancients' did as well – that gravitational attraction resulted from the direct and continuous action of God in the world. Like Kepler, who felt he had revealed God's geometrical blueprint, Newton considered himself chosen to restore ancient knowledge – and not just scientific knowledge. He spent years in theological and historical studies, believing that Christianity, like all other knowledge, had become corrupted over time, and endeavoured to restore its supposedly 'original' theology that did not include, for example, the divinity of Christ. He likewise laboured on ancient chronology, in part to get reliable reckoning dates for interpreting biblical prophecies about the end of the world. We return once again here to the broader, more inclusive view of natural philosophy relative to that of modern science. Newton saw 'the task of natural philosophy as the restoration of the knowledge of the complete system of the cosmos, including God as the creator and as the ever-present Agent'.

Chapter 4
The sublunar world

While many early modern natural philosophers looked up towards the heavens, even more looked anew at things on Earth. The sublunar world was the realm of the Earth and its four elements – earth, water, air, and fire – and the realm of change, of coming-to-be and of passing-away, a dynamic world of unceasing transformations. Heavy elements (earth and water) and heavy objects fell naturally towards the lowest point of the universe – its centre – where the Earth remained at rest. The light elements (air and fire) moved upwards towards the sphere of the Moon, the uppermost limit for the four elements. Thus each element found its 'natural place' in the scheme of things by means of a 'natural motion' based on its weight or levity. This Aristotelian system explained why rocks and rain fall downwards while smoke rises and a candle flame always points upwards. In the superlunar world, on the contrary, heavenly bodies were composed of the quintessence which, being neither heavy nor light, moved neither up nor down, but with eternal circular motion around the Earth. Early moderns re-examined the Earth, the elements, and the processes of change and motion, and formulated a range of systems for making sense of things. Some were expressly intended to replace the Aristotelian worldview, others tried only to refine it, and virtually none was completely free of Aristotle's influence. The result of observing, experimenting, and reconceptualizing the sublunar world was not the gradual formulation of a single worldview leading to the modern scientific perspective, but rather the creation of competing world systems that jostled for recognition and pre-eminence throughout the 17th century.

The Earth

Early modern natural philosophers considered the Earth, like the rest of the cosmos, to be only a few thousand years old. The chronology provided by the Bible, the oldest text available, drew the lineage of mankind back about 6,000 years. While only some readers interpreted Genesis 1 to describe a literal chronology involving six 24-hour days of creation (St Augustine had rejected such literalism in the 5th century), no one seriously thought that the Earth's prehuman history extended much further back in time. The largest estimates suggested a creation about 10,000 years old. This position was not a matter of dogma; there was simply no evidence to make one think otherwise. It was in the work of Niels Stensen (1638–86), better known by his Latinized name Nicholas Steno, that the idea of geological history emerged. Born in Denmark, Steno first applied himself to anatomy, becoming famous for his skill in dissection, with which he made important discoveries such as the salivary passage, known today as Stensen's duct. Like many other natural philosophers of his day, he toured European centres of learning, meeting other natural philosophers and exchanging new knowledge. In the 1660s, he settled in Florence under Medici patronage, and became interested in the layers of rock – what we call strata – visible in the Tuscan hills and the seashells found encased in them. He concluded that these layers must once have been soft mud laid down gradually by sedimentation, and therefore that lower strata must be older than higher ones. Wherever strata were not horizontal, he argued, they must have been disrupted by some upheaval after they hardened into stone. These conclusions did not cause Steno to revise estimates of the age of the Earth upwards – after all, mud can harden into brick relatively quickly – but they did indicate that the Earth's surface was subject to dramatic changes and that rocks preserve a record of these changes.

Towards century's end, several authors – especially in England – built upon Steno's work to compile 'histories of the Earth' to explain its current appearance. Most of them invoked global catastrophes as causal agencies and interleaved biblical and other historical accounts with natural philosophical ideas and observations. Thomas Burnet's *Sacred Theory of the Earth* (1680s) proposed six geological ages punctuated by cataclysmic biblical events. Edmond Halley and William Whiston (1667–1752), both associates of Newton, suggested that comets colliding with the Earth were major crafters of its history, causing such things as the inclination of the Earth's axis and Noah's flood.

Changes to the Earth's surface were studied first-hand by the Jesuit polymath Athanasius Kircher. While in Sicily in 1638, Kircher witnessed a violent earthquake and the eruption of Mt Etna. Vulcanism had not previously been a subject of study, in large part because Mt Vesuvius, the only active volcano on the European mainland, had been dormant for over three hundred years prior to its sudden and deadly eruption in 1631. Kircher travelled to observe the continuing eruption, and actually descended into the active crater to get a better look. He observed how volcanic action both destroyed old mountains and built new ones, dramatically altering the landscape. He attributed volcanic heat to the inflammation of sulphur, bitumen, and niter (a combination close to that of gunpowder) underground. Noting that the quantity of fire and molten rock emitted was too great to be produced within the mountain itself, he surmised that volcanoes must be vents for immense fires deep within the Earth. He thus concluded that the Earth cannot be merely 'pressed together from clay and mud after the Flood, hardly different from some lump of cheese', but had instead a complex and dynamic internal structure. He envisioned Earth's interior riddled with passages and chambers (Figure 8).

8. An idealized depiction of the hidden interior of the Earth and its volcanoes as envisioned by Athanasius Kircher, *Mundus subterraneus* (Amsterdam, 1665)

Some conveyed fire to volcanic vents from a fiery central core (he *never* literally conflated this core with Hell), while others allowed the passage of water, often from one sea to another. The flow of massive amounts of water through such passages generated ocean currents and turbulence. Collecting data from many sources, especially reports sent from Jesuit missionaries, Kircher compiled his encyclopedic *Subterranean World* (1665) containing, among much else, world maps showing ocean currents, volcanoes, and the possible locations of submarine passages.

In contrast to Kircher's observation of the Earth's most dramatic events, William Gilbert (1544–1603) performed quiet experiments at home to uncover another invisible feature of our planet. Gilbert, a physician to Elizabeth I, studied that ever-puzzling object, the magnet. His book *On the Magnet* (1600) surveys the properties of magnets, recounts experiments with them, and distinguishes magnetic attraction from the temporary ability of rubbed amber to attract straw. (For this latter phenomenon, he coined the word *electrical* – from the Greek *ēlectron* for amber.) Some of his experiments were inspired by those performed by Pierre de Maricourt in the 1260s, but Gilbert directed his studies towards a new goal. Pierre had used spherical magnets, or lodestones – pieces of the naturally magnetic mineral magnetite – and discovered that magnets have poles that he named north and south. Gilbert, also using spherical magnets, observed that iron needles placed on them mimicked exactly the behaviour of compass needles on the Earth. He thus concluded that the Earth is a gigantic magnet. It too has magnetic poles that attract the compass needle, just like a lodestone. (Previously it was thought that compasses pointed toward the *celestial* North Pole not towards a terrestrial pole.) In short, Gilbert used his spherical lodestone as a *model* of the Earth – reasoning by analogy, he extrapolated what he saw while experimenting with the lodestone to the whole Earth.

Gilbert's goal was to undergird Copernicanism, which had thrown the whole concept of natural place and natural motion into confusion. Putting the Earth in motion, spinning giddily on its axis and orbiting far above the centre of the universe, raised serious problems for physics. Why would heavy bodies fall to an Earth that is not at the centre? What caused the Earth to spin? Supporters of Copernicanism had to find a new physics that could reorder this chaos. Once Gilbert had argued that the Earth has magnetic poles, he stressed that those poles defined a real *physical* axis, and using the principle that everything in nature has a purpose, he argued that the purpose of that axis was to provide for the Earth's rotation. Furthermore, Earth's magnetic virtue animates it with intrinsic motive power; just as lodestones cause iron objects to move. The Earth's magnetic 'soul', as Gilbert calls it, not only causes compasses to turn north, but the planet to turn on its axis. Upon this foundation, Gilbert formulated a 'magnetical philosophy' wherein magnetic virtues permeate and govern the universe. Drawing upon the principle that like attracts like – the 'sympathy' of natural magic – the magnetic philosophy tried to solve the disruption of 'natural place' by suggesting that pieces of Earth are naturally attracted to the Earth, while pieces of the Moon are naturally attracted to the Moon. Thus earthly objects would fall towards the Earth regardless of the Earth's place in the cosmos. In Gilbert's universe, magnetic forces maintain order in both sub- and superlunar worlds, and his vision deeply influenced Kepler, Newton, and others.

Motion on Earth

While the magnetic philosophy tried to explain *why* bodies fall, Galileo endeavoured to describe mathematically *how* they fall. He built inclined planes, pendula, and other devices to study terrestrial motion. His *Two New Sciences* (1638), written while under house arrest, was the culmination of a study of motion he began in the 1590s. He discovered, contrary to Aristotle's claim, that all bodies fall at the same rate regardless of weight. With elegant logic he argued that if a ball rolled down an inclined plane speeds up and one rolled up an inclined plane slows down, then one rolled on a level surface – neither up nor down – would maintain a constant speed. Since on Earth that 'level' surface would actually be the curved surface of the globe, a ball rolled on its perfectly polished surface would circle it for ever. Using this 'thought experiment', Galileo both enunciated a principle of inertia (that moving bodies keep moving unless acted upon by an external agent), and brought the eternal circular motion of the heavens down to Earth – further eroding the distinction between sublunar and superlunar realms.

Methodologically, what Galileo ignored is as important as what he paid attention to. In describing motion, he never concerned himself with *what* is moving – a ball, an anvil, or a cow. In short, he ignored the *qualities* of bodies that Aristotelian physics emphasized. Galileo favoured instead their *quantities*, their mathematically abstractable properties. By stripping away an object's characteristics of shape, colour, and composition, Galileo gave idealized mathematical descriptions of its behaviour. A cold brown ball of oak doesn't fall any differently than a hot white cube of tin; Galileo reduces both objects to abstract, decontextualized entities able to be treated mathematically. A group known as the Oxford Calculators had begun applying mathematics to motion in the 1300s; in fact, Galileo begins his exposition of kinematics in the *Two New Sciences* with a theorem they enunciated. But Galileo went much further by linking mathematical abstraction tightly with experimental observation. As he conducted innumerable experiments, he sifted out air resistance and friction as 'imperfections' from the ideal mathematical behaviour that can be experienced only in thought. Plato, with his idea of a world that only imperfectly follows the eternal mathematical patterns according to which it was fashioned, might have found something to agree with in Galileo's perspective (even if Aristotle would have objected). Evoking the Christian image of the 'Book of Nature', Galileo wrote famously that 'this grand book, I mean the universe... is written in the language of mathematics, and its characters are triangles, circles, and other geometrical figures, without which it is humanly impossible to understand a single word of it'. The technique of reducing the physical world into mathematical abstractions, and eventually into formulas and algorithms, championed by Galileo, played a key role in producing a new physics, and stands as a distinctive feature of the Scientific Revolution.

Significantly, Galileo is content to *describe* motion mathematically without worrying about its *cause*. This feature of Galileo's work departs fundamentally from Aristotelian science where true knowledge is the knowledge of causes. Galileo's approach resembles an engineer's – a person more interested in describing and utilizing *what* an object does than *why*. Here Galileo draws upon his Northern Italian context where engineering and the learned engineer had achieved great prominence (see Chapter 6). The *Two New Sciences* makes the importance of practical engineering clear: its interlocutors meet amid construction works in Venice's shipyards, and discuss beam and tensile strength and scale-ups and scale-downs - topics of critical importance to engineers and architects. As a young professor in Padua, Galileo supplemented his meagre university salary by tutoring on mechanics and fortification. His later study of projectile motion - showing that projectiles follow a parabolic path - which we tend to remember primarily as a contribution to the physics of motion, continued earlier studies by Niccolò Tartaglia (1499–1557), a learned engineer who wrote his own *New Science* in 1537 about applying mathematics to motion, especially the motion of cannonballs, a topic of immediate practical importance for Italy's ever-warring states. It is easy to make the development of science too abstract and cerebral, and to forget that it is often driven by pressing and very practical concerns.

Water and air

The study of water for engineering purposes led to a sequence of important discoveries by Galileo's followers. His student and successor to his mathematics chair at Pisa, the Benedictine priest Benedetto Castelli (1577–1643), devoted himself to hydraulics and fluid dynamics - important practical questions in an era when Italy was awash with grand waterworks projects involving canals, fountains, irrigation, aqueducts, and sewers. The need to move water greater distances vertically (for example, out of deep wells or mines) led to the discovery that siphons could not draw water upwards to a height of more than about 34 feet. In the early 1640s, Gasparo Berti (c. 1600–43), a colleague of Castelli's at the University of Rome, tried an experiment to study this problem. With co-workers including Athanasius Kircher, he took a pipe 36 feet long and able to be closed at both ends, and mounted it vertically with its lower end in a basin of water

(Figure 9, left). He closed the bottom valve and filled the pipe completely with water. Then he closed the pipe at the top and opened it at the bottom. The water began to flow out, but stopped suddenly when the height of the column of water left in the pipe fell to 34 feet. What kept the water suspended at 34 feet – no higher and no lower?

Castelli's student Evangelista Torricelli (1608–47), who would later be given Galileo's position as mathematician and philosopher to the court of Ferdinando II de' Medici, devised a simple instrument analogous to Berti's pipe, but easier to handle. He took a glass tube about a yard long, sealed it at one end, and filled it with mercury. When the open end was inverted into a basin of mercury (Figure 9, right), the mercury in the tube began to drain out, but stopped when the column of mercury remaining in the tube was about 30 inches in height, about one-fourteenth the height at which the water had stopped in Berti's pipe. Significantly, mercury is about 14 times as dense as water – meaning that the height of any fluid left suspended in a tube was a direct function of the fluid's density. Drawing upon ideas of fluid equilibria worked out in earlier studies of water, Torricelli explained these results by saying that the weight of fluid left in the tube was balanced by the weight of the external air pressing down on the fluid in the basin. The idea that air had weight conflicted with Aristotle's system where it had none. Torricelli proposed not only that we 'live submerged at the bottom of a vast ocean of elemental air', but also that his instrument could measure and monitor changes in the weight of that air, leading to a new name for his device: the *barometer*, literally the 'measurer of weight'.

Some of the most celebrated experiments of the 17th century were designed to explore ideas provoked by Torricelli's tube. An elegant experiment to prove that it is the atmosphere's weight that keeps liquids suspended in the tube was proposed by the mathematician and theologian Blaise Pascal (1623–62), and carried out by his brother-in-law Florin Périer in 1647. Following

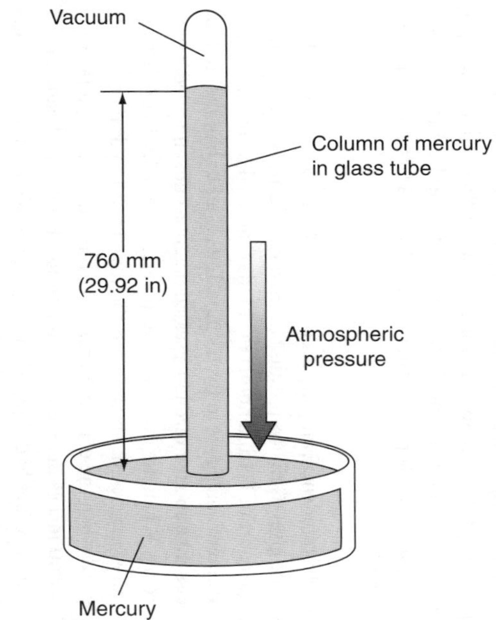

9. (left) Gasparo Berti's water barometer depicted in Gaspar Schott, *Technica curiosa* (Nuremberg, 1664); (right) A schematic of Evangelista Torricelli's simplified mercury barometer

Pascal's instructions, Périer prepared 'Torricellian tubes' in a monastery garden at the base of the Puy-de-Dôme, a mountain near their home in central France. He then carried one tube more than 3,000 feet up the mountain, where he found that the level of the mercury stood three inches lower. Upon returning down the mountain, the mercury regained its original height. At higher elevations, with less of the 'ocean of air' pressing down from above, the weight of air resting upon the mercury was reduced, and could therefore counterbalance less mercury in the tube.

A spectacular experiment performed before many spectators was that of the famed 'Magdeburg sphere' created by Otto von Guericke (1602–86), natural philosopher, mayor of Magdeburg, showman, and maker of wondrous devices. Von Guericke built two hemispherical copper shells with rims that fit smoothly together. He put them together to form a sphere, opened a valve installed on one half, and – using a device of his own invention modelled on a water pump – pumped the air out of the sphere. He closed the valve, and showed that teams of horses could then not separate the two halves because of the air's weight holding them together (Figure 10). Upon opening the valve, air rushed in, and von Guericke then easily separated the two halves with a flick of the wrist.

But what was in the space above the mercury or within von Guericke's sphere? Many experimenters believed it was literally *empty*, a vacuum – a highly controversial topic in the 17th century. Aristotelians and some others argued that a vacuum was impossible – as summarized in their catchphrase 'nature abhors a vacuum'. They saw the world as completely full of matter, a *plenum* – and some natural phenomena seemed to support them. They argued that the space contained air or some finer aerial substance drawn out of the mercury. Experiments endeavoured to resolve the point, but did not entirely settle the dispute between 'vacuists' and 'plenists'. Sound was not transmitted through the space, indicating that the air needed to carry sound had been removed. Yet light passed through – did not light, like sound, need

10. Otto von Guericke's showy demonstration that teams of horses could not pull apart the halves of a hollow sphere out of which the air had been pumped – evidence of the power of atmospheric pressure. Depicted in Gaspar Schott, *Technica curiosa* (Nuremberg, 1664)

some medium to transmit it? What are routinely seen as 'landmark' experiments in the history of science rarely proved as convincing to their contemporaries as they seem to moderns in retrospect. Experimenting, and especially interpreting results, is a tricky and contentious business, has always been and will always be so.

Robert Boyle (1627–91) soon joined the ranks of those studying air. As the youngest son of the richest man in Britain, Boyle had both time and resources to spend his life experimenting, mostly in his sister's house on London's Pall Mall, where he lived much of his adult life. He and several contemporaries noted the compressibility of air, specifically that the greater the pressure on a sample of air, the smaller its volume, a relation later called 'Boyle's Law' and still taught as such to chemistry students. In 1658, having heard of von Guericke's air pump, Boyle and the ingenious Robert Hooke built an improved version able to evacuate a large glass sphere, allowing various objects to be sealed up and observed as the air was pumped out (Figure 11).

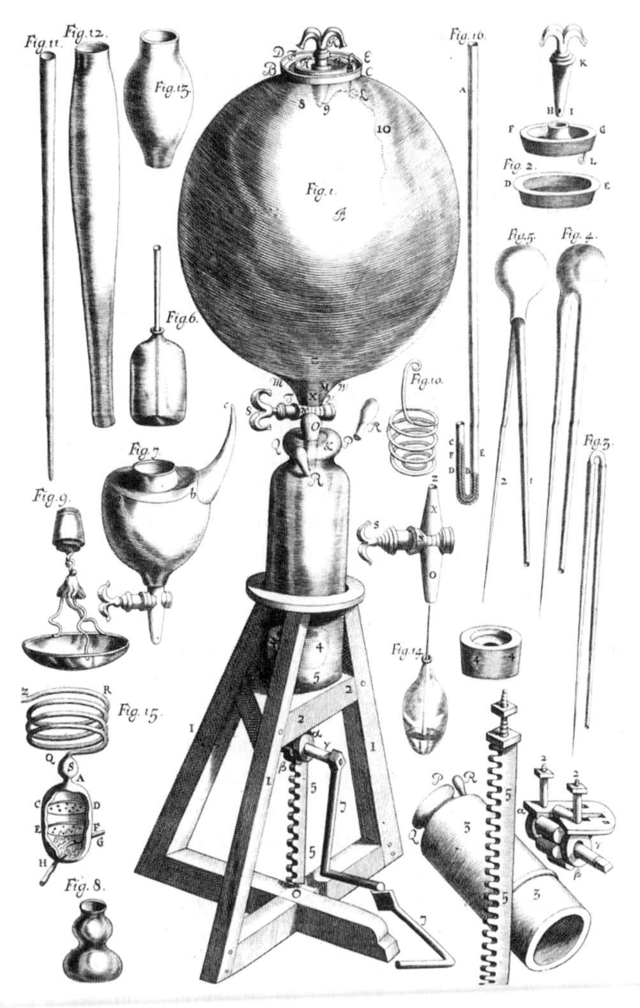

11. Robert Boyle's and Robert Hooke's air pump. Robert Boyle, ***New Experiments Physico-Mechanicall Touching the Spring of the Air*** **(Oxford, 1660)**

Boyle sealed a barometer (he probably coined this name for Torricelli's tube) in his air pump and watched the level of mercury drop as the air was withdrawn. He performed a dizzying array of experiments in the pump: from trying to ignite gunpowder, fire a pistol, or hear a watch ticking, to measuring how long various living creatures – cats, mice, birds, frogs, bees, caterpillars, and almost everything else – could survive without air. He also experimented with burning candles in the air pump, and noted the dependence of fire on the quantity of air available.

Fire: the chymists' tool

Long before the early modern period, the status of fire as a material element had been disputed. Amid such debates, one group regularly employed fire as their primary tool for studying and controlling matter and its transformations: the alchemists. The Scientific Revolution was alchemy's golden age. Today 'alchemy' is often taken to mean a single-minded (and futile) quest for making gold, something more or less 'magical' and thus distinct from chemistry. But in the early modern period, 'alchemy' and 'chemistry' referred to the same array of pursuits. Some historians today use the archaically spelled *chymistry* to refer to all these undifferentiated pursuits together. Gold-making, or *chrysopoeia*, was a key part of chymistry, but there was nothing 'magical' (in the modern sense) involved, simply a practice based on theories different from our own. Notebooks survive that record the daily operations of 'alchemists' and often reveal careful methodologies of experimental practice, textual interpretation, observation, and theory formulation. Besides the quest for gold, chymistry also included the broader study of matter and the production of articles of commerce such as pharmaceuticals, dyes, pigments, glass, salts, perfumes, and oils. The union of material production and natural philosophical speculation forms a central characteristic of this subject from its 4th-century origins in Hellenistic Egypt down to present-day chemistry.

The search for a method to turn lead into gold was not just wishful thinking. It rests upon the theory that metals are compound bodies produced underground by the combination of two ingredients called 'Mercury' and 'Sulphur'. When the two combine in the correct proportions and purity, they form gold. If there is not enough Sulphur, silver results. Too much Sulphur (a dry, flammable principle) produces iron or copper – their excess Sulphur can be demonstrated by the flammability, hardness, and the difficulty of melting these metals. Excess Mercury (a fluid principle) gives tin or lead – the soft, easily fusible metals. Thus transmutation was, in theory, a simple matter of adjusting these two components to the proportion found in gold. The observation that silver ores contained some gold and that lead ores contained some silver suggested that transmutation was occurring naturally underground, as poorly compounded metals were purified or 'matured' into more stable, better-concocted ones. The challenge was to effect this transformation artificially and faster. Chrysopoeians thus sought to prepare what they called the Philosophers' Stone, a material agent for bringing about transmutation. Once prepared in the laboratory, a small quantity of the Stone mixed with molten base metal was supposed to convert it into gold in a few minutes. Many texts claimed success in this process, and seekers after transmutation strove to replicate it. The difficulty lay in the intentional secrecy of such writings – the ingredients, process, and even the theory were hidden beneath codes, cover names, metaphors, and pictorial emblems, often of bizarre character (Figure 12).

Alchemy's secrecy arose in part from artisanal practices wherein it was necessary to preserve proprietary rights as trade secrets. Secrecy was encouraged by medieval laws forbidding transmutation out of fear of debasing the currency. But authors also justified secrecy by claiming that their knowledge was not only dangerous in the wrong hands, but also a privileged knowledge not to be divulged to those unworthy of it.

12. An alchemical allegory depicting the purification of gold and silver, a first step in making the Philosophers' Stone. The king represents gold, while the wolf jumping over the crucible (a vessel for refining metals) stands for the mineral stibnite, a material capable of reacting with and removing the silver and copper commonly alloyed with gold. The queen represents silver, and the old man (Saturn) lead, in reference to the process of cupellation, which uses lead to purify silver. From *Musaeum hermeticum* (Frankfurt, 1678)

The continuing British usage of 'chemist' to mean 'pharmacist' originated in the early modern period when most chymists devoted at least some of their efforts to making medicines. The application of chymistry to medicine began with the Provençal Franciscan friar Jean of Rupescissa (1310–c. 1362), who advocated the use of alcohol distilled from wine to prepare medicinal extracts. The use of chymistry to prepare medicinal substances expanded throughout the next century, before receiving its most vocal advocate in the larger-than-life figure of Theophrastus von Hohenheim, known as Paracelsus (1493–1541). Paracelsus criticized traditional medicine based on Greek, Roman, and Arabic authors and devised his own system based on a range of sources from direct observation to Germanic folk beliefs. He championed chymistry as the means to prepare virtually any substance into a powerful medicine, and showed little interest in chrysopoeia. His guiding idea was that noxious properties arise from impurities in otherwise wholesome substances, much like sin and death contaminated a world which, as God's creation, was intrinsically good. Using distillation, fermentation, and other laboratory operations, chymistry provided methods for dividing good from bad, medicine from poison. Paracelsus also taught that all substances were composed of three primary ingredients – Mercury, Sulphur, and Salt – a terrestrial trinity called the *tria prima* that mirrored the Divine Trinity and the triune nature of man – body, soul, and spirit. A process he called *spagyria* endeavoured to divide a substance into its *tria prima*, purify each, and then recombine them into an 'exalted' form of the original substance with enhanced medicinal power and no toxicity.

But Paracelsus went further: chymistry was not just a tool for making medicines, it was the key to understanding the universe. As Paracelsus' late 16th-century followers systematized his often chaotic writings (which, it was rumoured, he dictated when drunk), they formulated a chymical worldview that envisioned virtually everything as fundamentally chymical. The cycle of rain through sea, air, and land was a great distillation. The formation of minerals underground, the growth of plants, the generation of life forms, as well as the bodily functions of digestion, nutrition, respiration, and excretion were all seen as inherently chymical. God Himself became not the geometer of the Platonists, but the Master Chymist. His creation of an ordered world out of primordial chaos was akin to the chymist's extraction, purification, and elaboration of common materials into chymical products, and His final judgement of the world by fire like the chymist using fire to purge impurities from precious metals. This worldview saw even man's ultimate destiny as chymical. Upon death, the human soul and spirit separate from the body. The material body putrefies in the grave until, at the general resurrection, it is renewed and transformed, whereupon the purified soul and spirit are reinfused by God the Chymist to produce a glorified and eternal human being, just as in spagyria the *tria prima* are separated from a substance, purified, and recombined into a resynthesized and 'glorified' product.

Paracelsianism attracted many adherents. When Tycho first saw his nova in 1572, he had just stepped out of a laboratory where he was preparing Paracelsian remedies. He later built a laboratory into his observatory-castle in order to study what he called 'terrestrial astronomy', namely, chymistry ('as above, so below'). Because of the anti-establishment nature of Paracelsus' style, often expressed in rants against the Classical learning, universities, and licensed physicians, his ideas provoked heated debate and often found their greatest following among those outside of established circles. Indeed, chymistry as a whole lived most of its existence outside traditional halls of learning and suffered from an uneasy status. While physics and astronomy formed required parts of university study from the Middle Ages on, chymistry did not obtain an academic footing until the 18th century. One reason is that it could boast no Classical roots; neither Aristotle nor any other ancient authority wrote about it, unlike astronomy, physics, medicine, and the life sciences. Its close link to commerce and artisanal production, its practicality – and often messy, laborious, smelly character – further disabled it from being considered among respectable topics. Yet chymistry's emphasis on practical experiment also meant that it amassed a huge inventory of materials, knowledge of their properties, and facility for working with them. The commercial importance of this knowledge increased substantially throughout the 17th century and many chymists took an entrepreneurial route – sometimes engaged by princely or other patrons and mining operations to improve yields or seek transmutation, sometimes working independently to introduce new wares to the market place. Unfortunately, chymistry's ability to imitate gems and metals, and the claim of chrysopoeia to make gold, provided opportunities for fraud, leading to a widespread connection of chymistry with unscrupulous practices. Already in the late Middle Ages, Dante had put chymists – 'the apes of Nature' – into the eighth circle of Hell alongside counterfeiters and forgers, and later, 17th-century playwrights such as Ben Jonson in his *Alchemist* (1610) used the figure of the false chymist and his greedy clients to comic effect.

Most 17th-century training in chymistry took place in medical contexts. In Germany, Johannes Hartmann (1568–1631) became the first professor of *chemiatria* (chymical medicine) in 1609. His appointment was made at the University of Marburg, a Calvinist institution newly established (and hence more able to be innovative) by Moritz of Hessen-Kassel, a prince whose court supported chrysopoeians, Paracelsians, and other chymists. In France, regular chymistry instruction began at the Jardin du Roi in Paris, a botanical garden founded to propagate and study medicinal plants. A succession of lecturers at the Jardin gave 'how to' courses based on laboratory demonstrations that were open to the public. Private lecturers, often pharmacists, also offered courses of chymistry, such as Nicolas Lemery who taught from his house in Paris. His textbook *Cours de chymie* (1675) became a best-seller. Indeed, the dozens of chymical textbooks published in France and Germany established a didactic tradition that compensated for chymistry's absence from university curricula.

Chymistry's practical flavour does not mean that it did not contribute significantly to natural philosophical theories – quite the opposite. One of the most important developments of the 17th century, the re-emergence of atomism, was built in part upon chymical ideas and observations. Already in the late 13th century, the Latin alchemist known as Geber used a quasi-particulate matter theory to explain chemical properties. He explained, for example, gold's density and resistance to corrosion by theorizing that its 'tiniest parts' were tightly packed together leaving no space between them. Iron was more loosely packed, leaving spaces that rendered the metal lighter in weight and providing places for fire and corrosives to enter the metal and break it apart into rust. Later chymists continued to develop the idea of stable, minute particles, and to use it for explaining their observations. Mainstream Aristotelians often rejected such notions, for they claimed that substances lose their identity when combined. But practising chymists knew that they could often recover starting materials at the end of a sequence of transformations. For example, chymists knew that silver treated with acid 'disappears' into a clear, homogeneous liquid that passes freely through filter paper. When treated with salt, that liquid precipitates a heavy white powder, and that powder when mixed with charcoal and heated to red heat, gives the silver back again in its original weight. This well-known experiment indicated that the silver maintained its identity throughout, despite appearances and despite having been broken into invisibly small particles able to pass through the pores in paper. Chymical operations provided the best evidence for such 'atoms'.

Atomism and mechanism

The chymical tradition of particulate conceptions of matter cross-pollinated with a revival of ancient atomism. Ancient Greek atomism began with Leucippus and Democritus in the 5th century BC. They conceived of a material world composed of indivisible atoms moving in void space, their coming together and moving apart in ever-changing combinations gave rise to all the changes we see. Their conceptions largely died out in antiquity. Aristotle refuted them at length, and although Epicurus (341–270 BC) made atomism foundational for his moral philosophy, when Epicureanism fell out of favour because of its tendencies to atheism and hedonism (Epicurus intended neither consequence), atomism went out with the bathwater. A revival occurred only after the rediscovery of Lucretius' poem *On the Nature of Things*, a Roman popularization of Epicurus, in 1417. But Lucretius' emphasis on the link between atomism and atheism initially rendered his book unpalatable. Ironically, the rehabilitation of Epicurean atomism was due to a priest, Pierre Gassendi (1592–1655). Gassendi denied that atoms are eternal (only God is eternal) and that they move of their own accord (God set them moving), argued for the immateriality and immortality of the human soul, and then built a comprehensive world system using invisible particles and their motions as its fundamental explanatory principle. His system and others like it came to be called the 'mechanical philosophy'.

The mechanical philosophy holds that all sensible qualities and phenomena result from the size, shape, and motion of invisibly small pieces of matter – variously called atoms, corpuscles, or simply particles. Strict mechanical philosophers maintained that there is only one sort of 'stuff' out of which everything is made, and that only the differing shapes, sizes, and motions of the tiny particles of this single element provide the variety of substances and properties we perceive. Coherent with his disregard for qualities in favour of quantities, Galileo argued that most qualities, like hot and cold, colours, odours, and tastes do not actually exist, but are only the result of how minute particles affect our sense organs. For Galileo, and for later mechanical philosophers, the only real qualities – the *primary* qualities – were the size, shape, and mobility of particles. All other qualities were *secondary*, having existence only in the sensor, not the sensed. For the mechanist, vinegar seems sour only because its sharp and pointy particles prick the tongue. Apart from the tongue, 'sourness' has no meaning. A rose appears red only because of the way its particles modify reflected light and the way that modified light acts upon our eyes. The rose's pleasant smell results from an effluvium of particles the flower emits, that travel through the air into our nose where they strike the olfactory organ, producing motions which, when conveyed to the brain, are converted into a sensation of smell. This viewpoint fundamentally opposes Aristotelian ways of viewing the world, wherein sensible qualities have real existence in objects, and play a crucial role in explaining the object's nature and effects.

This system was mechanical in two senses. First, effects were caused only by mechanical contact – like a hammer on stone, or billiard balls colliding. There is no room for action-at-a-distance or powers of sympathy. Second, the world and objects in it – even plants and animals in the widely influential mechanical philosophy of Descartes – were conceptualized as *machines*. Mechanical philosophers compared the world to a complex clockwork – like the huge mechanical clocks of the period where hidden gears, weights, pulleys, and levers caused visible hands to turn, bells to ring, figurines to dance and bow, and mechanical roosters to crow, all in perfect order and regularity. The term 'machine of the world' (*machina mundi*) dates back to Lucretius and was used in the Middle Ages to express the complex regularity of the universe, but for those authors *machina* meant something more like frame or fabric, and expressed the interdependence of the various parts of creation. Mechanical philosophers however, gave the image the sense of an automaton, that is, something artificial but imitating the actions of a living thing mechanically. Mechanical perspectives reflected the increased technological prowess of the day, and shifted conceptualizations of the world away from living biological models towards lifeless machinery. This viewpoint led even to a reconceptualization of God Himself. Rather than a geometer, chymist, or architect, God was seen increasingly as a mechanic or watchmaker – a technician who designed and assembled the world machine. This image, which became particularly entrenched in late 17th-century England, forms the ultimate background to modern-day discussions of 'intelligent design'. In the early modern period, when theology and natural philosophy shaded seamlessly into one another, scientific and religious concepts grew and developed hand in hand, each one affecting and responding to the other.

As mechanical philosophers strove to apply their principles to all natural phenomena, one particular challenge was to explain the 'hidden qualities', sympathies, and actions-at-a-distance that had frustrated Aristotelians and formed the basis of natural magic. Mechanists' favoured solution was an appeal to invisible material effluvia – 'steams' of particles that carried effects from one body to another. For example, fire can heat an object at a distance because rapidly moving fire-particles emanate from the flame and strike the object. Other explanations required more inventive solutions. Descartes explained magnetic attraction by suggesting that magnets emitted a constant stream of screw-shaped particles. Iron, he postulated, contains screw-shaped pores, hence, the particles emitted by the magnet enter iron's pores and turn in them, thereby 'screwing' the iron closer to the magnet. Even the reflex action of turning away from a gory sight was explained on the basis of an efflux of sharp particles that wound the eyes.

Robert Boyle not only gave the mechanical philosophy its name, but joined it to chymistry in particular, recognizing chymistry's special ability to reveal the workings of the world. Boyle pursued all four major aspects of 17th-century chymistry: chrysopoeia, medicine, commerce, and natural philosophy. He sought avidly for the secret of making the Philosophers' Stone and tried to contact 'secret adepts' who could offer assistance. He claimed to have witnessed the Stone's use and tested the gold that he saw it produce from lead, and was responsible for having an English law forbidding transmutation repealed in 1689. He collected new chymical medicines, especially less expensive ones valuable for the relief of the poor (medical care and pharmaceuticals were overpriced then just as today). He also advocated the application of chymistry towards useful ends, for the improvement of trades, commerce, and manufacture. Perhaps most famously, he promoted chymistry as the best means for studying the world, and strove to elevate chymistry's status. Boyle explained that he devoted himself to chymistry, which his friends considered 'an empty and deceitful study', because it provided the best evidence for the particulate systems proposed by mechanical philosophers. As an example, he showed experimentally how saltpetre could yield both a fixed alkaline salt and a volatile acidic liquid, and how combining the two regenerated the saltpetre. The conclusion he drew was that compound substances could be taken apart into pieces and the pieces put back together to reform the original substance, just like parts of a machine. Although Boyle rejected much of Paracelsianism, such 'reintegrations' (as he called them) bear a striking resemblance to *spagyria*, and indeed, Boyle built his ideas upon a long foregoing tradition of both chrysopoeia and chemiatria.

The mechanical philosophy waned by the end of the 17th century. Boyle himself became less enthusiastic about it as he realized its overextension could lead to determinism, materialism, and atheism. If the world were only an array of colliding particles, there would remain no room for free will or divine providence. If God is a clockmaker, did He start the world running and then abandon it, or must He regularly readjust it as if He were less than a master mechanic? Chymists remained unimpressed by a strict mechanical philosophy – the vast array of properties they saw everyday did not seem explicable by the lean notions of a single kind of matter with differently shaped particles. Life processes were likewise far too complex for simple mechanics to explain beyond a certain point. Finally, Newton's forces of attraction, a kind of action-at-a-distance, were not reducible to mechanical explanation. The triumph of Newtonianism in fact meant the defeat of strict mechanism.

Evolving Aristotelianism

Aristotle and Aristotelianism have come in for quite a few knocks throughout this chapter. Indeed, one interpretation of the Scientific Revolution is that it was all about the rejection of a moribund Scholastic Aristotelianism. But this view fails to acknowledge the flexibility and continuing evolution of Scholasticism. While proponents of various 'new' philosophies of the 17th century routinely caricatured and criticized Aristotelianism with harsh rhetoric, other natural philosophers remained within that 'Aristotelian' framework and continued to update the system and to work productively. Neither in the late Middle Ages nor in the early modern period did being 'Aristotelian' or 'Scholastic' mean holding stubbornly to every claim made by Aristotle himself. Even Aristotle's own greatest student, Theophrastus, continued the Aristotelian tradition by disagreeing with his master on several points. In the Middle Ages, natural philosophers universally cited Aristotle, but often simply as a starting-point for their own explorations which frequently came to conclusions contrary to Aristotle's. By the Renaissance, there were many different and even conflicting Aristotelianisms.

Experimental and mathematical approaches to natural philosophy were not key parts of Aristotle's own work, but they increasingly became so for 17th-century Aristotelians. The Jesuits provide the clearest example of an explicit commitment to maintaining an Aristotelian natural philosophy, yet many, like Riccioli and Grimaldi, carried out extensive experiments relating to Galilean kinematics, and incorporated ideas and findings expressly contrary to Aristotle. Similarly, Niccolò Cabeo (1586–1650) rejected Gilbert's pro-Copernican interpretation of his magnetic experiments, but Cabeo's own experiments with the magnet were as extensive. By the end of the century, Jesuits had adopted many of the particulate and mechanical views expounded by Gassendi and Descartes, but within an 'Aristotelian' framework. For its proponents, Scholasticism remained a useful and flexible *method* of proceeding in the study of nature not necessarily a body of conclusions. While retaining a conservative stance towards the many innovations of the 17th century, they were nevertheless active participants and contributors to the Scientific Revolution.

What certainly did happen in the Scientific Revolution is that Aristotelianism acquired serious and radically different competitors, something it had not encountered in the late Middle Ages. Throughout the early modern period, new worldviews – the magnetical, chymical, mathematical, natural magical, mechanical, and others – emerged as challengers and plausible alternatives, while Scholasticism endeavoured to incorporate new material and ideas within an 'Aristotelian' framework. The continuing arguments between defenders of the various world systems resulted not only in a wealth of polemical pyrotechnics but also in a broad range of eclectic responses to the pressing challenge of establishing a new, and preferably comprehensive philosophy of nature. From our modern perspective, it is hard to imagine the broad diversity of viewpoints and approaches in regard to fundamental questions and methods that flourished in the early modern period, or the fertility and fervency with which an ever-increasing number of natural philosophers explored their world and devised systems – some small, some vast – to try to make sense of it all. This is one of the important ways in which the period of the 16th and 17th centuries was in fact 'revolutionary'.

Chapter 5
The microcosm and the living world

In addition to the world beyond the Moon and the world beneath the Moon, there was a third world that riveted the attention of early modern thinkers: the *microcosm* or 'little world' of the human body. Early modern physicians, anatomists, chymists, mechanists, and others focused on this living world that we inhabit. They explored its hidden structures, endeavoured to understand its functions, and hoped to find new ways of maintaining its health. The life that characterizes the human body naturally connects it to the rest of life on Earth – its flora and fauna. The catalogue of these living creatures exploded during the Scientific Revolution, thanks not only to voyages of exploration but also to the invention of the microscope, which revealed unimagined worlds of complexity in ordinary objects and new worlds of life within a drop of water.

Medicine

The human body was the first concern of the physician, and medicine had a high profile both socially and intellectually throughout the High Middle Ages and the early modern period. Alongside law and theology, medicine formed one of the three higher faculties of the university. The medical knowledge taught in 1500 was an accumulation of medieval Arabic and Latin experience and innovation built upon a core of ancient Greek and Roman teachings. Galen, Hippocrates, and Ibn Sīnā (or Avicenna,

c. 980–1037) stood as its chief authorities, and humoral theory formed its foundations. Humoral theory maintained that bodily health depended not only upon the proper functioning of the various organs, but also upon a balance, called *temperament*, among four 'humours', or fluids, found in the body: blood, phlegm, yellow bile, and black bile. These four humours corresponded with the four Aristotelian elements and shared their pairings of primary qualities (Figure 13).

The physician's role was to assist nature in re-establishing humoral balance by prescribing particular diets, daily regimens, and medicines. This predominantly Galenic medicine worked by 'contrary cures', that is, if a patient has (what we still Galenically call) a 'cold', resulting from excess phlegm (the cold and wet humour), then hot and dry foods and medicines should be administered to help restore balance. For a fever, cold and wet medicines are needed, cold baths, or perhaps bleeding to withdraw excess blood and its hot quality.

The many relationships held to exist between the superlunar world and the human body beautifully illustrates the connectedness of the early modern world. The macrocosm's influence on the microcosm was largely unquestioned, even if the details of this interaction were constantly debated. Thus astrology played a key role in both diagnosis and treatment; medicine, not prognostication, was probably astrology's chief application. Each bodily organ corresponded with a zodiacal sign and was particularly susceptible to influences from the planet that resembled it in qualities (Figure 14).

The brain, for example, a cold and wet organ, is influenced most by the Moon, a cold and wet planet. (Hence, someone with disordered brains is still today called a lunatic – from *luna*, Latin for Moon – or more colloquially, 'moony'.) Knowing the planetary positions at the onset of an illness could therefore assist in diagnosis by helping the physician understand prevailing

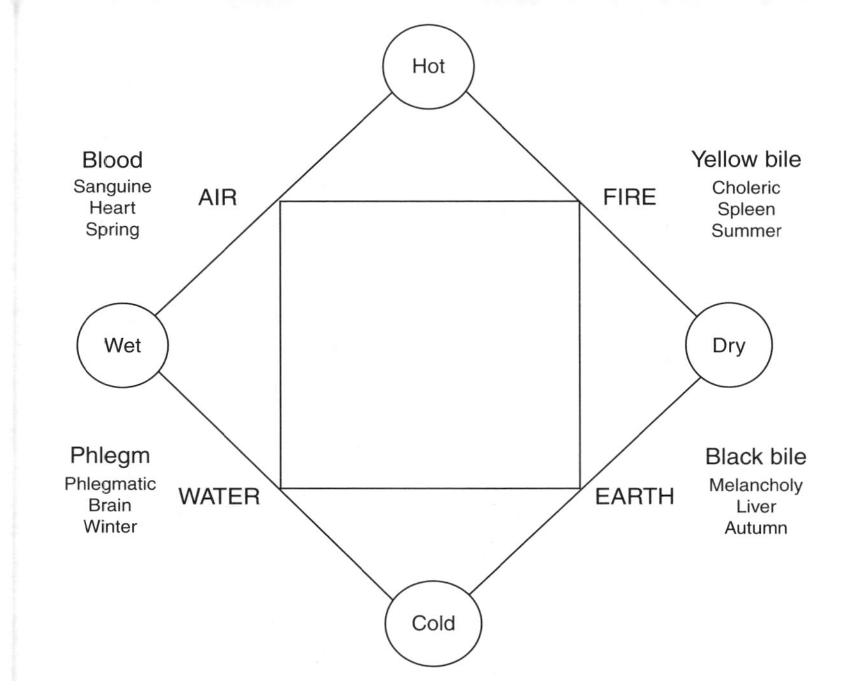

13. A 'square of the elements' showing their qualities and their relationships with the four humours, four bodily complexions, and four seasons

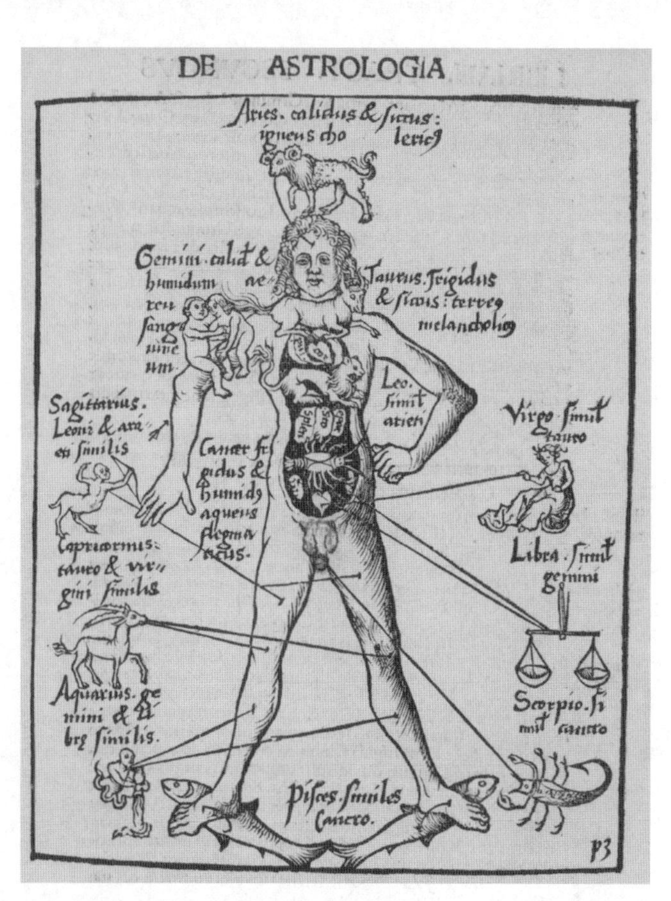

14. A chart of the organs and their zodiacal correspondences. From the early modern encyclopaedia compiled by Gregor Reisch, entitled *Margarita philosophica* (Freiburg, 1503)

environmental influences or localize potentially affected parts of the body. Furthermore, each person was held to have a unique ratio of humours – called his *complexion* – imprinted at birth by the then-prevailing planetary influences; this means that every patient must be restored to his own particular complexion.

One size does not fit all in early modern medicine. Medical treatments had to be tailored to each patient; the same pill could not be used on everyone, and a particular diet and regimen had to be followed in parallel with treatment. A physician might therefore examine the patient's natal chart to gain insight about the patient's complexion. Astrological calculations could also assist in timing medical treatments, according to the Hippocratic idea of 'critical days', namely, that during the progress of an illness there are points of 'crisis' that must be successfully overcome for the patient to recover. Diagnosis also relied on the examination of urine – portable reference charts provided tables of the colour, smell, consistency, or even taste of patients' urine and the relation of these indicators to various ailments. The same is true of the rate, rhythm, and strength of the pulse.

Methods of medical treatment, at least among licensed physicians, did not change dramatically during the Scientific Revolution. Despite a slow evolution in response to new ideas and discoveries, the core of Galenic and Hippocratic medicine continued well into the 18th century (although astrological diagnoses began to wane in the 17th). This endurance reflects both the stability of medical school curricula and the guild or licensing structure of medicine that promoted conservatism. Innovations thus often came from outside the body of licensed physicians. The strict licensing of physicians was, however, possible only in large urban centres. In most places, a variety of healers with little or no formal medical education attended to people's health and far outnumbered licensed physicians. Virtually every householder kept a list of home remedies for family and neighbours. Apothecaries made both simple and prepared medicines easily available such that virtually anyone could compound even exotic (and sometimes dangerous) medicines. Surgeries were carried out by barber-surgeons, a group with lower status and less formal training than physicians. 'Empirics', unlicensed physicians offering a variety of medicaments and treatments, found their best trade in the cities despite frequent attempts to ban them from London, Paris, and other major centres. In dramatic contrast to modern medical practice, some treatments were done contractually, that is, the physician's remuneration was dependent upon his success.

Radically new medical approaches such as Paracelsianism and, in the 17th century, Helmontianism were taken up more avidly by unlicensed practitioners, often in direct challenge to the medical establishment. Nevertheless, new chymical approaches to medicine made slow but steady ingress into official pharmacopiae and the practices of professional institutions like the Royal College of Physicians of London, established in 1518. In France, the conservative Galenic faculty of medicine in Paris and the pro-Paracelsian faculty at Montpellier waged a decades-long battle over the risks and rewards of chymical medicines. This conflict also reflected fault lines running between the royal, centralized, and predominantly Catholic Parisians, and the provincial, mostly Protestant Montpellians. Their most fervent debate, over the medical use of antimony - a toxic mineral - came to an end only after 1658 when Louis XIV, having fallen ill during a military campaign and not responding to traditional treatments by the royal physicians, was cured by a vomit induced by a dose of antimony in wine administered by a local physician. The Parisian medical faculty thereafter had little recourse but to vote to legalize the use of this Paracelsian *vin émetique*.

Anatomy

Anatomy witnessed significant development in the early modern period. Although Galen stressed the importance of anatomy in antiquity, Romans considered the violation of dead bodies by dissection socially and morally unacceptable, and thus Galen dissected apes and dogs and transferred his findings by analogy to human beings. (Nevertheless, he undoubtedly saw exposed human innards from time to time during his position as a physician to gladiators.) Only in Egypt were human dissections carried out in antiquity, probably because opening the body and removing its organs was already familiar there due to the practice of mummification. During the late Middle Ages, however, human dissection became standard in Italian medical schools such as Padua and Bologna. By about 1300, medical students were required to observe a human dissection as part of their training. There is no basis whatsoever to the 19th-century myth that the Catholic Church prohibited human dissection. Human dissection was hampered mostly by a shortage of corpses. Since respectable people would not permit their bodies or those of their kin to be displayed and cut up before an audience, dissections were dependent upon the availability of corpses from executed criminals, often foreigners.

Interest in human anatomy increased greatly in the early 16th century, particularly in Italy, culminating in Andreas Vesalius' (1514–64) monumental work *On the Structure of the Human Body*, published in 1543, the same year as Copernicus's *On the Revolutions*. Born in Flanders, Vesalius trained at Padua and became lecturer of surgery there the day after receiving his MD. Assisted by a judge who timed executions conveniently (without refrigeration or preservatives, corpses had to be dissected immediately), Vesalius performed many careful dissections, noting the errors of Galen and other authors, and grouping parts of the human body in new ways, no longer just functionally but structurally as well. Drawing upon the skills of artists from Titian's workshop, Vesalius supervised the production of detailed anatomical drawings, and these formed a main feature of his book, whose text explained each illustration and anatomical feature in great detail. Producing so richly illustrated a book would have been impossible without the printing press. Still, the lavish volume was expensive, spurring Vesalius to produce a cheaper version for students, through which his ideas, discoveries, and organizing principles gained wide circulation. Increased interest in anatomy led to the construction of anatomy theatres, first in Padua (1594), then in Leiden (1596), Bologna (1637), and elsewhere. Although intended for teaching medical students, these theatres, especially those in Northern Europe, attracted large audiences of interested (or fashionable) onlookers from the wider public as well.

Dissections were not restricted to either human corpses or medical schools. With the rise of 17th-century scientific societies, animal dissections became a major part of their activities. In the 1670s and 1680s, the young Parisian Royal Academy of Sciences received the bodies of exotic animals that had died in Louis XIV's menagerie, including an ostrich, lion, chameleon, gazelle, beaver, and camel. While dissecting the last of these, the head of the Academy, Claude Perrault (1613–88), nicked himself with the scalpel and died from the resulting infection. In the 1650s and 1660s at Oxford, and then at the Royal Society in London, several workers dissected not only dead but living animals, especially dogs, in experiments too gruesome for the modern reader to stomach (Boyle himself was disturbed by them). These experiments endeavoured to learn the actual workings of nerves, tendons, lungs, veins, and arteries. Often, they included the injection of various fluids to observe their movement through the body and their physiological effects, as well as blood transfusions, sometimes from one species into another, including attempts to cure sick human beings with blood transfused directly from healthy sheep.

This interest in blood and the movement of bodily fluids stemmed in part from William Harvey's (1578–1657) arguments for the circulation of the blood published in 1628. According to Galen, the venous and arterial systems are separate units. The liver continuously produces dark venous blood that the veins distribute through the body as nutriment. A portion of this blood is drawn into the heart, where it passes through pores in the tissue, or septum, dividing the right and left ventricles. There, air drawn from the lungs via the pulmonary artery converts it into bright arterial blood, which then nourishes the body through the arterial system. No blood ever returns to the heart. The 16th-century anatomists, however, found problems with Galen's system. They questioned the existence of pores in the septum, and found that the pulmonary artery was full of blood, not air. The latter observation led to the proposal of the 'lesser circulation': venous blood passes from the heart through the lungs, then returns to the heart before flowing out into the body. At the University of Padua, Harvey studied with the greatest anatomists of the day, notably Girolamo Fabrizio d'Acquapendente (1537–1619), who had described 'valves' he found in the veins. Harvey later remarked that this discovery led him to consider a wider circulation of the blood.

Harvey noted that the volume of blood pumped by the heart would exhaust the body's supply within moments unless it was somehow recirculated. Using ligatures to stop the flow of blood selectively, he experimentally deduced the 'greater circulation', namely, that the heart pumps blood circularly through connected arterial and venous systems. Harvey found the blood's circular motion satisfying, since it meant that the microcosm mimicked the macrocosmic heavens whose natural circular motion Aristotle considered the most perfect. Indeed, Harvey maintained Aristotelian approaches and methods, and focused attention on the heart and blood partly because of the central role Aristotle had given them – another example of Aristotle's continuing importance in the Scientific Revolution. Harvey was unable, however, to detect the tiny capillaries that connect arteries to veins. These structures were first seen only four years after Harvey's death by Marcello Malpighi (1628–94), who observed the movement of blood through minute vessels linking the pulmonary vein to the pulmonary artery in the transparent lung tissues of frogs; he extrapolated that similar vessels connected arteries to veins throughout the body. To make this observation, Malpighi used a relatively new invention: the microscope.

Microscopy, mechanism, and generation

The origins of the microscope in the early 17th century are obscure, but like its sister the telescope, it revealed a new world and provoked new ideas. Galileo used a device similar to his telescope to magnify small objects, but the first drawings made using the microscope appear in studies of bees carried out in 1625 by Francesco Stelluti and Federico Cesi and dedicated to Pope Urban VIII, whose Barberini family used the bee as an emblem. In the 1660s, Robert Hooke built an improved microscope to examine everything from tiny insects like lice, to frost crystals, and the fine structure of cork, which he found divided into chambers he called 'cells' after their resemblance to monastic living quarters. Antoni van Leeuwenhoek (1632–1723), a draper and surveyor in Delft, devised the simplest and most powerful magnifiers. He built more than five hundred microscopes using a tiny spherical glass bead as their single lens, and published more microscopical observations than any other author. He subjected an incredible array of objects to his microscopes, observing 'worms' in human and animal semen, corpuscles in blood (and their movement through capillaries in the tail of a young eel), bacteria in dental plaque, and swarming 'animalcules' in pond water and infusions of vegetable matter. His discovery of spermatozoa fed into a lively debate over the nature of animal and plant generation. Leeuwenhoek himself supported *preformationism*, the idea that a tiny version of new offspring was contained within each spermatozoon, or, according to some of his contemporaries, within each egg. The opposite view, *epigenesis*, held that embryonic structure was produced *de novo* and in successive stages during gestation. Preformationism appealed especially to mechanical philosophers because it reduced generation to a simple matter of mechanical growth – a tiny organism simply got bigger by assimilating new matter. As such it abandoned the immaterial vital forces most epigenesists considered necessary for crafting amorphous material – semen and/or menstrual blood or the fluid of an egg – into an organized and differentiated embryo. Harvey, an epigenesist, by opening chicken eggs at various stages of their development, observed that blood formed first, which he took as evidence that it was the seat of life and of a vital soul that guided the formation of the offspring. Preformationism however provoked the question of where and when the tiny form of the new organism actually began. A few suggested that all future generations were contained, one inside the next, within the first of a species created by God.

The microscope's revelation of seemingly mechanical structures in living bodies excited mechanists in particular, and accordingly most microscopists of the late 17th century were mechanists. They embraced Harvey's circulation of the blood in part because it characterized the heart as a pump – a mechanical device, although Harvey was far from a mechanist himself – and they strove to reduce complex living systems to mechanical principles. In Florence, Giovanni Alfonso Borelli (1608–79), for example, analysed animal motion in terms of simple machines – conceptualizing bones, tendons, and muscles as levers, fulcrums, and ropes, and bodily fluids and vessels as hydraulics and plumbing, thus launching what has come to be called biomechanics. In London, Nehemiah Grew (1641–1712) explored the hidden anatomical structures of plants, helping to establish plant physiology. Some mechanists even hoped that improved microscopes would allow the direct observation of atoms, their shapes, and their motions, exposing to direct observation the fundamental explanatory principles of the mechanical philosophy.

Microscopical observations, like all others, were open to conflicting interpretations. While the discovery of spermatozoa could be interpreted to favour preformationism, the contemporaneous discovery that countless living creatures appeared on their own in stale water strongly favoured established notions of spontaneous generation – that living creatures could emerge from non-living material – which in turn favoured the epigenic idea that living structures emerge from originally amorphous matter. For centuries previously, most natural philosophers had assumed that simple life forms appeared spontaneously under certain circumstances – a rotting bull carcass generated bees, mud generated worms, putrefying flesh generated maggots. In a series of famous experiments in the 1660s carried out at the Medici court,

Francesco Redi (1626–97) left samples of meat out to rot, some covered with a mesh or cloth and others in the open air. Those in the open air produced maggots, while none appeared when access by flies was prevented. As in most cases of experiments seen retrospectively as 'definitive', Redi's experiments did not immediately stamp out belief in spontaneous generation since other explanations of the results could be (and were) offered, and Redi himself allowed that some insects – like the oak gall wasp – might be produced directly from plant matter. Although moderns routinely scoff at belief in spontaneous generation, it is worth pointing out that any modern scientist who does not believe in a special creation of the first life form by God's miraculous intervention must consequently believe in the spontaneous generation of life from non-living matter.

Neither the microscope nor the mechanical view of living systems lived up to expectations. The limits of magnification and resolution, given the material and optical systems available, were soon reached. Microscopical investigation had revealed such enormous complexity in living systems that mechanistic explanations seemed increasingly inadequate to account for either their formation or their functioning. Yet even while mechanical approaches enjoyed their greatest popularity, more vitalistic models also flourished. In fact, the divide between non-living and living was not at all clear-cut in the 17th century, and many thinkers hybridized mechanical and vitalistic systems. For example, few mechanists were so rigid that they denied the existence of an animating soul in living systems. Such a soul need not be like the immaterial, immortal human soul of Christian theology, but rather was considered to exist in various forms or levels in various entities (for modern readers perhaps the term 'vital spirit' better expresses the concept). These notions date back to Aristotle, who had proposed three levels of soul: a *vegetative* soul in plants responsible for overseeing growth and the assimilation of nutrition; in animals, a further *sensitive* soul to govern sensation and movement; and in human beings, in addition to the vegetative and sensitive soul, a *rational* soul to govern thought and reason. For many, while mechanical principles could explain particular bodily functions and structures, the organization and maintenance of the organism as a whole – not to mention consciousness and awareness – were functions of soul.

Helmontianism

Perhaps the most comprehensive new system of medicine to emerge in the 17th century was that of the Flemish nobleman, physician, chymist, and natural philosopher Joan Baptista van Helmont (1579–1644). Van Helmont combined chymistry, medicine, theology, experiment, and practical experience into a cohesive and highly influential system. His autobiographical statements express a dissatisfaction with traditional learning and a desire to pursue new knowledge that is typical of Scientific Revolution-era thinkers. He recounts how he attended the University of Louvain, but refused his degree because he felt he had learned nothing. He then studied with the Jesuits and felt no better off. Then he obtained an MD, but finding the foundations of medicine 'rotten', he turned to Paracelsianism, only to reject much of that as well. Thus van Helmont endeavoured to start afresh, calling himself a 'philosopher by fire', meaning that his training came not from traditional learning but from experiments in chymical furnaces. Indeed, van Helmont was an extraordinary observationalist, describing the origin, symptoms, and progress of several maladies that were not otherwise recognized until centuries later.

Van Helmont rejected the four Aristotelian elements and the Paracelsian *tria prima*, claiming instead that water was the single underlying element of everything. Not only did this idea harken back to the oldest-known Greek philosopher Thales, but more importantly (for van Helmont) also to Genesis 1:2 where the Spirit of God brings forth the world by 'brooding [like a hen] upon the *waters*'. The Belgian philosopher sought experimental confirmation of this idea, most famously by planting a five-pound willow sapling in two hundred pounds of soil, and watering it for five years. At the end of that time, the tree weighed 164 pounds but the soil had lost scarcely any weight; therefore, van Helmont concluded, the entire composition of the tree must have been produced from water alone. According to van Helmont, *semina* (seeds) implanted in the world at creation have the power to transform water into all substances. These *semina* are not physical seeds like a bean, but immaterial organizing principles, like the invisible vital principle that organizes the fluid of an egg-yolk into a chick. Fire and putrefaction destroy the *semina* and their organizational power, thus turning substances into air-like substances van Helmont called 'Gas' (from the word *chaos*, and the direct source of our word for the third state of matter). Thus burning charcoal and fermenting beer release a choking *Gas sylvestris*, and burning sulphur a stinking *Gas sulphuris*. In the cold parts of the atmosphere, this *Gas* finishes converting back into primordial water and falls as rain, thus closing the cycle of water's successive transformations in van Helmont's economy of nature.

Similar to, but more sophisticated than Paracelsus, van Helmont held that bodily processes were fundamentally chemical. He recognized the acidity of gastric juice responsible for digestion, and performed analyses of bodily fluids – especially of urine to find the cause and cure of one of the 17th century's most painful maladies, kidney and bladder stones. Yet chemical processes could not suffice on their own to explain life processes; they had to be directed by a quasi-spiritual entity lodged in the body and called the *archeus*. For van Helmont, the *archeus* regulates and governs bodily functions. Sickness results from a weakened *archeus*, unable to perform its duties; medical treatment must therefore work to strengthen the *archeus*. Accordingly, van Helmont rejected Galenic notions of complexion, the four humours, and methods of healing. Diseases like the plague, he said, are not due to humoral imbalance but to external 'seeds' of disease invading the body and transforming it. A strong *archeus* can dispel these seeds, but a weak one needs help. (Note that in both Galenic and Helmontian medicine, the physician's role is always to *assist* natural processes, never to divert them or to assert control over the body.) Van Helmont also emphasized the role of the patient's mental and emotional state, and claimed that the power of imagination can cause physical changes in the body. Helmontian ideas deeply influenced many physicians, physiologists, and chymists.

Mechanist and vitalist conceptions of living systems were not irreconcilable but rather two ends of a continuum; many physicians and natural philosophers embraced intermediate positions. Like his contemporary van Helmont, Gassendi invoked seeds as powerful principles able to organize matter into new forms. But while van Helmont's seeds were immaterial, Gassendi's were special combinations of physical atoms (divinely organized) that acted mechanically on matter. Indeed, mechanist and vitalist speculations produced hybrid medical systems in the 18th century, such as that of Georg Ernst Stahl (1659–1734) which emphasized the mechanical nature of chemical transformations but the need for vital powers to organize and govern living systems. Herman Boerhaave (1668–1738), perhaps the most influential voice in 18th-century medicine, especially in pedagogy, drew together diverse strands of 17th-century natural philosophy. As professor of medicine and chemistry at Leiden University's medical school, Boerhaave strongly advocated both Hippocratic methods of healing (emphasizing environment and patient individuality) and the importance of chemistry for medical education. His approach to medicine and the body combined aspects of Boyle's mechanical philosophy, Newton's physics, and van Helmont's 'seeds'. Boerhaave's reforms of medical education were adopted throughout much of Europe (hence he was sometimes called the 'Teacher of Europe'), and proved foundational for significant changes that would occur in 18th-century medicine.

Plants and animals

The study of flora and fauna – what we would call botany and zoology – expanded enormously in the 16th and 17th centuries. The traditional textual location for such material was an encyclopedia tradition, stemming from the massive *Natural History* compiled by Pliny the Elder (23–79 AD) in his attempt to collect and popularize Greek learning for a general Roman audience. Encyclopedic accounts of plants and animals filled medieval herbals and bestiaries, and this format continued into the Scientific Revolution. One of the most famous is the five-volume *History of Animals* by Conrad Gessner (1516–65) with its hundreds of woodcuts. Many such volumes would, however, appear strange to modern readers, for they blend naturalistic and descriptive details about various species with a mass of literary, etymological, biblical, moral, mythological, and metaphorical meanings that had accumulated around each animal or plant since antiquity. No account of the peacock would be complete without mention of its pride, of the serpent without its deceitful role in Adam's Fall, of the plantain (a common plant that grows in footpaths) without reference to how it signifies the well-trodden way of Christ. Plants and animals are not presented as isolated species, but rather within rich networks of meaning and allusion. They are both natural objects and emblems dependent upon the vision of a world of layered meanings, a world simultaneously literal *and* metaphorical, a world full of symbolic messages to be read. As a result, even fabulous animals such as the unicorn, dragon, and various monsters are described alongside well-known creatures, not necessarily because the authors believed they roamed the Earth, but more because whether or not they existed in the physical world, they nevertheless carried meaning thanks to their existence in the literary world. While modern readers might consider such texts 'quaint', credulous, or encumbered with 'non-scientific trivia', their original audience would probably consider modern botanical or zoological descriptive texts sterile and oddly disengaged from humanity.

Two developments of the early modern period diverted this emblematic tradition into other directions. The needs of medicine to identify herbal remedies was the first. As humanist scholars continued to revive, edit, and publish Greek medical texts, it became increasingly necessary to identify the medicinal plants these texts mentioned and to help locate them in the wild. Hence there was a demand for new herbals that bridged the gap between ancient texts and what grew in 16th-century fields. To accomplish this task, new herbals not only linked common names with ancient Greek ones, but provided accurate, naturalistic illustrations of them. Just as Vesalius collaborated with artists from Titian's workshop, so too a new generation of 16th-century botanists worked with artists to produce herbals with extensive illustrations drawn from life, such as Otto Brunfels' *Living Images of Plants* (1530–6) and Leonhart Fuchs' *History of Plants* (1542). The second development was the expansion of European horizons. On the narrowest level, ancient authorities like Dioscorides described mostly Mediterranean plants and did not recognize Northern European species, hence it became necessary to provide accounts of plants that did not have a Classical pedigree. The same problem, but on a much larger scale, existed in terms of the countless plants and animals encountered for the first time in voyages outside of Europe, especially in the Americas. Food plants like potatoes, corn, and tomatoes, medicinal plants like 'Jesuit's bark' (cinchona, the source of quinine, a cure for malaria), and new animals like the opossum, jaguar, and armadillo greatly increased the catalogue of flora and fauna known to Europeans. These new arrivals had no accumulated networks of correspondence and emblematics, and so could not be fit into the traditional format of herbals and bestiaries. In Spain, where most reports from the New World first arrived, those charged by the king with organizing the information were forced to give up established encyclopaedic methods based on Classical models like Pliny not only because new findings rendered old categories obsolete, but also because the unrelenting flow of new information made it impossible to organize the knowledge comprehensively.

Spaniards in the New World, often members of religious orders, struggled to chronicle native plants, animals, and medical practices, sometimes collaborating with indigenous scholars to produce illustrated texts. José de Acosta (1539–1600), sometimes called the 'Pliny of the New World', was a Jesuit in Peru who, besides founding five colleges, wrote a natural history of Latin America that was widely published, translated, and referenced in Europe. In 1570, King Philip II sent his physician Francisco Hernández on an expedition specifically to seek out New World medicinal plants. Hernández spent seven years, mostly in Mexico, cataloguing plants and inquiring about their properties from indigenous healers, while a team of native artists produced the illustrations for a six-volume *Plants and Animals of New Spain* (it describes about 3,000 plants and dozens of animals). Frustrated by the impossibility of inserting new plants into Classical classification schemes, Hernández even adopted native names to create a new botanical taxonomy. Meanwhile, the Franciscan friar Bernardino de Sahagún (1499–1590), working with Aztec assistants and informants at the Colegio de Santa Cruz at Tlatelolco in Mexico, produced the *General History of Things in New Spain*, a lengthy work in both Spanish and Nahuatl describing Aztec culture, customs, society, and language. At home in Spain, the physician Nicolás Monardes (1493–1588) compiled a *Medicinal History of Things Brought Back from Our West Indies* that described dozens of New World species. Portuguese scholars such as Garcia de Orta (1501–68) and Cristóvao da Costa (1515–94) likewise reported on their findings of medicinal plants and new animals in India and elsewhere in South and East Asia.

The search for new medicines drove the study of new plants, and consequently the establishment of botanical gardens, usually in the context of medical schools. Medicinal gardens had been a part of monasteries throughout the Middle Ages, and new botanical gardens built upon this foundation and expanded it for pedagogical and research purposes. The first botanical gardens opened in Italy at the universities of Pisa and Padua in the 1540s and Bologna in 1568, along with the establishment of professorial chairs in medical botany.

Other centres of medical instruction followed suit – Valencia (1567), Leiden (1577), Leipzig (1579), Paris (1597), Montpellier (1598), Oxford (1621), to name a few. These gardens were laid out in strict order, with species grouped by therapeutic property, morphology, or geographical origin. Seeds, roots, cuttings, and bulbs were sought after, traded, and exchanged, thereby expanding the range of plants available in gardens across Europe. The interest in unusual plant cultivation and hybridization spread to private individuals, leading to the celebrated 17th-century 'tulipomania' in the Netherlands, where newly made bourgeois fortunes were drained to acquire rare hybrids, and artists preserved exotic flowers in still-lifes.

A widespread interest in the exotic and rare expressed itself in collecting natural historical specimens of all kinds in 'cabinets of curiosities' (Figure 15). While these collections were in one sense

15. The cabinet of curiosities of Ole Worm from *Museum Wormianum (The Wormian Museum, or, A History of the Rarer Things both Natural and Artificial, Domestic and Exotic, which the author collected in his house in Copenhagen)* (Leiden, 1655)

the forerunners of museums, they also functioned to display the power, wealth, connections, and interests of their collectors and to invoke wonder at the marvels of nature and art. Princes and noblemen as well as scholars amassed collections that included both *naturalia* – botanical, zoological, and mineralogical specimens – and *artificialia* – mechanical contrivances, stunning works of art and craft, and ethnographic and archaeological objects. Ulisse Aldrovandi (1522–1605) compiled one of the earliest such collections (part of which still survives in Bologna), and a guided tour by Athanasius Kircher of his museum at the Collegio Romano was a 'must see' for 17th-century visitors to the Eternal City. The physical arrangements of the objects within the space of the cabinet emphasized the connections between objects – often ones that we would not consider. Thus these cabinets became microcosms of another sort, displaying and emblematizing the diverse, the marvellous, and the exotic of the linked worlds of man and nature all compressed into a single room.

Chapter 6
Building a world of science

Science is more than the study and accumulation of knowledge about the natural world. From the late Middle Ages down to our own day, scientific knowledge has been used increasingly to change that world, to give human beings greater power over it, and to create the new worlds in which we now live so much of our lives, seemingly ever more separated from the natural world. More and more people today become so surrounded by the world of artifice constructed by technology that they notice their dependence upon it only when it fails, and then find themselves as helpless as a medieval farmer when the rain does not fall on his crops. Thus moderns often react with consternation when the natural world reasserts itself by intruding inconveniently upon this artificial world – when meteorites or solar flares knock out satellite communications, lightning strikes cut off electrical power, or volcanic eruptions shut down airline traffic. The proliferation of technology has changed the daily world of human beings more radically than anything else in the last few centuries. That explosion of technology simultaneously depends upon and encourages scientific inquiry. The 16th and 17th centuries witnessed a special turn towards applying scientific study and knowledge to address contemporaneous problems and needs.

The world of artifice

In Renaissance Italy, ambitious new engineering projects transformed landscape and cityscape. Canals and waterworks claimed new land and provided drinking water and transportation routes. Filippo Brunelleschi's (1377–1446) immense double-shelled dome for the cathedral, with its innovative construction techniques set a new skyline for Florence. New urban design fulfilled the humanist emphasis on civic life and proclaimed the wisdom and power of ruling princes while new fortifications protected their interests. As is often the case, one new technology drove the development of others. The technological transformation of warfare in the 15th century by the increasing use of gunpowder and production of portable bronze cannons rendered medieval fortifications obsolete – their soaring battlements provided excellent targets for artillery. Thus a new system of fortification had to be developed. New designs for fortification drew upon geometrical principles and became standard parts of a nobleman's education. Pressing practical concerns (and princely ambitions) produced, first in 16th-century Italy, then elsewhere, a class of learned engineers and architects who, following the lead of the ancient models Archimedes and Vitruvius, increasingly turned to mathematical principles and analysis to solve practical problems. Falling between artisans who relied on accumulated manual experience and scholars removed from practical affairs, this emergent class provided a crucial background for the increasing deployment of mathematics to investigations of the world, an essential feature of the Scientific Revolution. Leonardo da Vinci (1452–1519) is one early example of this 'intermediate' group, as is the military engineer Tartaglia in the mid-16th century. At the end of the century Galileo drew inspiration and borrowed methods from the learned engineers.

Both practicality and the humanist desire to emulate the ancients inspired renovations of the city of Rome. Papally sponsored projects explored and rebuilt ancient acqueducts and sewers.

The dilapidated 4th-century St Peter's was pulled down to build the immense new basilica that stands there today, and provoked one of the most spectacular engineering feats of the 16th century: the moving of the Vatican obelisk. A single stone the height of a six-storey building and weighing over 360 tons, the obelisk had been erected by the Romans in the 1st century. In 1585, as the new St Peter's encroached upon the obelisk, Pope Sixtus V issued a call for proposals to move the ancient Egyptian stone to a new location – the first time an obelisk had been moved in 1,500 years. The engineer Domenico Fontana (1543–1607) won the commission. Using the combined power of 75 horses and 900 men operating 40 windlasses, five 50-foot-long levers, and eight miles of rope, Fontana successfully lifted the monolith – encased in iron armatures – straight up off its base on 30 April 1586. The operation was considered so important that the Pope allowed part of the newly completed basilica to be torn down in order to allow optimal operation of the levers and windlasses. Fontana then lowered the obelisk onto a carriage (Figure 16), transported it along a causeway, and re-erected it where it stands today, the focal point of St Peter's Square.

Renaissance achievements, and the economic and military engines that supported them, required materials. Accordingly, the period 1460 to 1550 witnessed a mining boom, particularly in central Europe where mineral resources were richest. Medieval mining had been largely a small-scale operation that exploited surface deposits. But the demands of early modern Europe – iron and copper for arms and artillery, silver and gold for coinage – drove more organized, larger-scale mining and the development of better smelting and refining techniques. Deeper shafts and increased scales required more mechanization – water wheels to drive bellows and rock-crushing equipment, pumps to drain mines and ventilate shafts – as well as greater organization of labour. Perhaps the most famous writer on mining, Georgius Agricola (1494–1555), a German humanist and educator, endeavoured to organize and promote mining knowledge. His massive and richly illustrated

16. Moving the Vatican obelisk, from Domenico Fontana, *Della trasportazione dell'obelisco vaticano* (Rome, 1590)

Latin treatise *On Metallic Things* endeavoured to ennoble an otherwise dirty enterprise by linking German mining practices with Classical literature and creating a Latin vocabulary for metallurgy. The landscape of felled trees, smoke, and streams of runoff shown incidentally in Agricola's illustrations underlines how such technological growth came with increasing costs to the environment. Probably more useful to actual practitioners were the German-language books of Lazar Ercker (c. 1530–94), an overseer of mining operations. His books are filled with practical experience about treating ores, assaying metals, and preparing chemical products like acids and salts including saltpetre, the crucial ingredient in gunpowder. By the mid-16th century, the boom was over – ended as much by the depletion of European mines as by the flood of metals from the New World that depressed metal prices, making the working of European mines less profitable.

The promise of the New World spurred developments in cartography and navigation. Late medieval navigational charts, or portolans, indicated only coastlines overlaid with rosettes of compass-headings from particular points. These charts were useful for relatively short journeys in the Mediterranean or along coastlines, but not for providing a geographical perspective or for journeying across oceans. Ptolemy's 2nd-century *Geography*, rediscovered in the fifteenth century, described using a grid of east–west and north–south lines (latitude and longitude, respectively) for mapping. Late 15th-century maps – such as Waldseemüller's – adopted this method, employing curved latitude lines and longitude lines that converged towards the poles. The Flemish cartographer Gerhardus Mercator (1512–94) popularized the now more familiar Mercator projection, where parallel longitude lines intersect straight latitude lines at right angles. Although it distorts land masses at high latitudes, this method of projecting the spherical Earth on a flat map was easier for navigation (at least at low latitudes) and was favoured by Spanish cosmographers and navigators.

The compass and quadrant – instruments to determine heading and latitude, respectively – had been used for navigation since the Middle Ages, but there existed no reliable method to determine longitude. This inability did not present a serious problem while vessels stayed in European waters or within sight of land. But crossing oceans was a perilous venture without accurate longitude measurements. Since locating a place requires both latitude and longitude, the lack of longitude presented so serious a problem for cartographers and navigators that finding a method to determine it became the most urgent technological problem of the period. Competing seafaring states – Spain, the Netherlands, France, and England – offered rich prizes for anyone who could devise a reliable method.

Time-telling is the key to longitude. Every hour of difference in local time between two places translates into fifteen degrees of longitude in separation (hence a modern 'time zone' is roughly fifteen degrees wide). But how to know the time at two distant locales simultaneously? One could take along a clock set at the ship's place of origin, and compare its reading with the time at the ship's location as determined by observations of Sun or stars. Unfortunately, early modern clocks were barely reliable to twenty minutes a day. Galileo's observation that pendula beat out a constant tempo regardless of the amplitude of their swing suggested a new regulator for time-keeping. He began designing a pendulum-regulated clock while under house arrest, but never built it. It was Christiaan Huygens in the Netherlands who produced the first workable pendulum clock in 1656, resulting in a huge leap in reliability – at least for land-based clocks. On a rocking ship, pendulum clocks did not run accurately. Thereafter, Huygens and Robert Hooke experimented independently with spring-powered clocks, but these too proved insufficiently accurate aboard ships. Still, Hooke's study of springs led to his enunciation of the relationship between the extension and the force of a spring, known today as 'Hooke's Law', just as Huygens's work led to refinements of the laws of simple harmonic motion. (The longitude problem itself was solved only in the 18th century using innovative chronometers devised by the English instrument-maker John Harrison that could keep accurate time even at sea.)

The alternative to a manmade clock was a celestial one – some astronomical event whose time of occurrence at a reference site could be calculated and then compared with the local time of the event at the observer's site. Spanish cosmographers of the 16th century successfully used coordinated observations of lunar eclipses to determine the longitude of settlements in the Spanish empire, but lunar eclipses are too rare for navigation. Jupiter's four moons, however, undergo more frequent eclipses – the innermost satellite Io has an eclipse every forty-two hours – and Galileo proposed using them as time-keepers. The astronomer Gian Domenico Cassini (1625–1712) explored this idea most fully, and in the 1660s compiled timetables of these eclipses. But once again, while this system worked on land – it was used successfully for correcting terrestrial maps – it proved impractical to observe the eclipses telescopically from a moving ship. Nevertheless, while testing the idea, observers noticed that some eclipses occurred several minutes later than predicted. Realizing that this discrepancy was greatest when Jupiter was farthest from Earth, the Danish natural philosopher Ole Roemer (1644–1710) proposed in 1676 that light has a finite speed – the eclipse's apparent delay was due to light's travel time across space – and made possible a rough measure of its speed.

These few examples indicate how technological application and scientific discovery were inextricably linked; each drove and enhanced the other. The notion of a 'pure' versus an 'applied' science does not apply in the 17th century – if it applies anywhere. Minimizing the importance of practical needs – whether military, economic, industrial, medical, or sociopolitical – as the driving force behind developments of the Scientific Revolution would yield an artificial and erroneous depiction of how things really happened.

The linkage of scientific discovery to practical application is perhaps most often associated with Sir Francis Bacon (1561–1626). Born into a well-placed family, educated as a lawyer, elected to Parliament, ennobled as Lord Verulam, and eventually named Lord Chancellor of England (and ousted on bribery charges), Bacon lived most of his life in the halls of power. Accordingly, the topic of power and the building of empire was rarely far from his thoughts. He asserted that natural philosophical knowledge should be *used*; it promised power for the good of mankind and the state. He characterized – or caricatured – the natural philosophy of his day as barren, its methods and goals misguided, its practitioners busy with words but neglecting works. Indeed, although Bacon expressed scepticism of natural magic's metaphysical foundations, he praised magic because it 'proposes to recall natural philosophy from a miscellany of speculations to a magnitude of works'. Natural philosophy should be *operative* not speculative – it should do things, make things, and give human beings power. He considered printing, the compass, and gunpowder – all technological achievements – to have been the most transformative forces in human history. As a result, Bacon called for nothing less than a 'total reconstruction of sciences, arts, and all human knowledge'.

Methodology is crucial to Bacon's desired reform. He advocated the compilation of 'natural histories', vast collections of observations of phenomena whether spontaneously occurring or the result of human experimentation, what he called forcing nature out of her usual course. After sufficient raw materials had been collected, natural philosophers could fit them together to formulate increasingly universal principles by a process of induction. The key was to avoid premature theorizing, navel-gazing speculations, and the building of grand explanatory systems. Once the more general principles of nature had been uncovered, they should then be used productively. Yet Bacon did not advocate a crass utilitarianism. Experiments were useful not only when they produced fruit (practical application) but also when they brought light to the mind. True knowledge of nature served both for 'the glory of the Creator and the relief of man's estate'. While Bacon is clear that one goal of his enterprise is to strengthen and expand Britain – although neither Elizabeth I nor James I responded to his petitions for state support of his ideas for reform – on a larger scale Bacon saw the goal of such operative knowledge as to regain the power and human dominion over nature bestowed by God in Genesis, but lost with Adam's Fall.

Crucially, Bacon considered not only the methods and goals of natural philosophy but also its institutional and social structure. He insisted that older ideals of solitary scholarship had to be replaced with cooperative, communal activity. Indeed, his programme of fact-collecting would require enormous labours, and although he embarked upon such collections himself, he was able to complete very little. Towards the end of his life, he cast his vision of reformed natural philosophy, and the improved society it could create, into a Utopian fable entitled *The New Atlantis* (1626). The story describes the island of Bensalem, a peaceful, tolerant, self-sufficient, Christian kingdom in the Pacific. The happy state of this island is due not only to wise kingship, but even more to the work of Solomon's House, a state-sponsored institution for the study of nature devoted to 'the knowledge of causes and the secret motions of things; and the enlarging of the bounds of Human Empire, to the effecting of all things possible'. The members of Solomon's House study nature communally, although with division of labour and hierarchical arrangement – lower levels collect materials, middle levels experiment and direct, and the highest levels interpret. In Bensalem, Baconian natural philosophers form an honoured and privileged social class, supported by the government, and in service to state and society. Bacon's vision proved inspirational for many 17th-century natural philosophers across Europe as they negotiated their own shifting positions within society.

The rise of scientific societies

Today, scientific research takes place at many sites, some of which even bear resemblances to one or more features of Solomon's House. Scientists work in universities, in governmental, industrial, and independent laboratories, at sites of large and unique instruments (like telescopes or particle accelerators), out in the field or at research stations and outposts, in zoos, museums, and elsewhere. Individual scientists are bound together into social groups by professional organizations, scientific societies and academies, research teams, correspondence, and most recently, the Internet. Funding for scientific research comes from government research grants, corporate research and development, universities, and private philantropy. These three features – physical place, social space, and patronage – are essential to the functioning of modern science. The establishment of these features during the Scientific Revolution was essential for constructing the world of science we know today. Throughout the 17th century and into the 18th, the work of natural philosophers became increasingly formalized. Individuals banded together into private associations which in turn evolved into national academies of science. Individual exchanges of information by letter grew into printed journals. Self-funded amateur and university-based natural philosophers were joined by the first salaried professionals.

During the late Middle Ages, natural philosophical inquiry took place predominantly in universities, monastic settings, and – to a much lesser extent – a few princely courts. These traditional loci of activity remained important during the 16th and 17th centuries, but were supplemented by new venues. Essential to the humanist movement of the Renaissance was the establishment of learned circles of scholarship outside the universities. Within these circles, scholars shared their work with like-minded individuals, receiving support, critique, and recognition as well as occasional patronage. These early groups were mostly literary or philosophical in character. By the late 16th century, however, natural philosophers had expanded the model, thus giving rise to the first scientific societies. The earliest such societies were established in Italy, where dozens were founded in the 17th century – more than anywhere else in Europe. Most of them, however, remained local and short-lived.

One of the earliest societies was the Accademia dei Lincei (Academy of Lynxes). Its name alludes to the emblematic character of the lynx as sharp-eyed and perceptive. The Academy was founded in Rome in 1603 by Prince Federico Cesi – then an 18-year-old Roman nobleman – and three companions, and functioned for about 30 years. Cesi founded his academy upon the belief that the investigation of nature was a complex and laborious affair that required group effort. There were never more than a handful of Linceans, but they included the advocate of natural magic Giambattista Della Porta, Galileo, and Johann Schreck, later a Jesuit missionary who brought European scientific knowledge to China. The Linceans pursued projects in all branches of natural philosophy, often independently but occasionally collaboratively, such as their long-term endeavour to publish the *Treasury of Medicines from New Spain* (1651) from manuscripts of Francisco Hernández's expedition to Mexico that had been brought to Italy from Spain. The Linceans embraced the new chemical approaches to medicine, promoted Galileo's work (his 1613 *Sunspot Letters* and 1623 *Assayer* were published under the auspices of the Lincei), and performed microscope studies. Cesi's early death in 1630 robbed the Lincei of its leader and patron, and precipitated its collapse.

In 1657, the Accademia del Cimento was founded at the Medici court in Florence, due in large part to Prince Leopoldo de' Medici's personal interests in natural philosophy. Its motto *Provando e reprovando* ('By testing and retesting') encapsulates the group's focus on performing experiments. The Medici court provided a central locale for communal studies, something the

Lincei had lacked, and Medici patronage kept it running financially. Many members were followers of Galileo, and the group continued several of his research projects and methods. Nevertheless, members of this Florentine Academy worked on everything from anatomy and the life sciences to mathematics and astronomy, and paid special attention to studies and improvements of new instruments such as the barometer and thermometer, in which Leopoldo himself participated. The work of Redi, Malpighi, Borelli, and of many other notable Italian natural philosophers, was carried out within the Cimento. Disagreements between members, the departure of several luminaries, and Leopoldo's nomination as a cardinal which required him to spend more time in Rome, led to the Cimento's closure in 1667. In its decade of existence, the Cimento established the most visible exemplar of a voluntary association of natural philosophers devoting themselves communally to the experimental investigation of nature.

By mid-century, scientific societies spread north of the Alps. In 1652, four physicians in Germany formed the *Academia naturae curiosorum*. Throughout its early years, this 'Academy of Inquirers into Nature' focused mostly on medical and chemical topics. The academy's statutes, published in 1662, declare as its goals 'the glory of God, the enlightenment of the art of healing, and the benefit resulting therefrom for our fellow men'. It grew rapidly, and although members lived widely dispersed throughout German-speaking lands and thus could not meet regularly as a corporate body, the Academy served to link them together virtually, especially through the annual publication (beginning in 1672) of a volume of collected papers submitted by the members. In 1677, Holy Roman Emperor Leopold I gave it official recognition. The 17th-century foundation expanded well beyond medical and life sciences in succeeding years, and eventually developed into the present-day German National Academy of Sciences Leopoldina.

At Oxford University in the 1650s, a group known simply as the 'Experimental Philosophy Club' began meeting at Wadham College to discuss natural philosophy, experiment with mechanical devices, and observe dissections and demonstrations. Christopher Wren and Robert Hooke were early members, and they were joined by Robert Boyle and other notables of mid-century England. Following the Restoration of Charles II in 1660, several members of this Club joined with others to draw up statutes for a more formal corporate organization, and received royal charter in 1662 as the Royal Society of London for the Improvement of Natural Knowledge. The Royal Society, in continuous existence to the present day, marks a new stage in the evolution of scientific societies. Like the Cimento (with which it maintained correspondence), the communal performance of experiments was central, but the Royal Society was envisioned as a much larger, more formalized organization. Over 200 Fellows were soon elected, although most choices among the English nobility reflected wishful thinking about financial rather than any intellectual contributions. Explicitly taking Bacon and his prescriptions as their model, the Royal Society envisioned public and social aims for itself. Indeed, the Royal Society can be seen as an attempt to realize Solomon's House. Many of the early Fellows had been involved in Utopian, educational, and entrepreneurial schemes during the civil war years, and brought these aims to the Society. They strictly avoided sectarian and political attachments, hoping to find in natural philosophy a basis for agreement that could overcome the factionalization of the civil war years immediately preceding.

The Fellows held regular meetings at Gresham College in London, where experiments were performed and new results and observations presented. Virtually every notable British natural philosopher of the period (and since) was a Fellow. The Society's membership soon reached beyond British borders, and election as a Fellow, then as now, carried substantial prestige. Perhaps the most important innovation linked with its early years was the establishment in 1665 of the *Philosophical Transactions*, the first scientific journal, by the Society's secretary Henry Oldenburg. Started initially as Oldenburg's private endeavour – he vainly hoped to earn a living from subscriptions – the *Transactions* soon became conceptually linked with the Royal Society although formally so only later. Oldenburg maintained a vast correspondence (as a result of which he was once imprisoned in the Tower as a spy), and could thus report scientific goings-on across Europe. The *Philosophical Transactions* published not only the Royal Society's activities, but reports and scientific letters from abroad, as well as book reviews. Despite publication predominantly in English, it became a crucial organ for European scientific life – a place to publish observations, announce findings, establish priority, and conduct disputes. Newton's papers on light, optics, and his new telescope appeared there, as did van Leeuwenhoek's microscopal observations mailed in from Holland, and Malpighi's anatomical studies sent from Italy. Arguments over comets jostled for room with reports of monstrous births, and issues appeared whenever Boyle had something relatively brief to report.

Despite its ambitions, the Royal Society suffered the problems common to early scientific societies – loss of key members, financial woes, and lack of patronage. Many of its grand schemes came to naught as a result. A majority of Fellows were inactive and paid their dues sporadically or not at all, and the Crown's sole gift to the Society was the adjective 'Royal'. Its Baconian project for the improvement of trades floundered on the understandable unwillingness of tradesmen to share their proprietary expertise. The English response outside of natural philosophical circles was no better – the Society, its Fellows and activities, were lampooned on the stage in Thomas Shadwell's *Virtuoso* (1676) and their claims to public utility acidly parodied by the 'Voyage to Laputa' in Jonathan Swift's *Gulliver's Travels* (1726). Oldenburg's death in 1677 caused the *Philosophical Transactions* to lapse for a time, and Boyle's in 1691 meant the loss of the Society's most active and generous Fellow. Newton, a Fellow since 1672, became President of the Society in

1703, by which time he was recognized as England's pre-eminent natural philosopher. His prestige breathed new life into the organization, but his tendency to favour work that promoted his own narrowed the former breadth of the Society's activities. Nevertheless, the Society became securely established by the middle of the 18th century, and has carried on ever since.

Unlike the Royal Society that was established from the bottom up, the Parisian Académie Royale des Sciences was established from the top down. It was the brainchild of Jean-Baptiste Colbert (1618–83), finance minister to Louis XIV. Colbert intended both to add glory to the Sun King as patron of arts and sciences, and to centralize scientific activity in ways useful to the state – part of the larger centralization of France that characterized Louis' long reign. The Académie held its first meeting in 1666, with twenty academicians headed by Christiaan Huygens, who had been recruited from Holland. They met twice a week at the King's Library, were expected to work communally (this did not always go smoothly), and received a salary and research support. The French thus realized Bacon's vision far better than did his countrymen. In return for royal funding, academicians were expected to find scientific solutions to state problems. It is no coincidence that the two most highly paid members – Huygens and Cassini – were brought to France while working on the crucial problem of longitude. Academicians also tested water quality at Versailles and throughout France, evaluated new projects and inventions, examined books and patents, solved technical problems at the royal printing press and elsewhere, and produced the first accurate survey of France. The last enterprise, by finding France to be smaller than previously thought, is said to have led Louis XIV to quip that his own academicians had succeeded, where all his enemies had failed, in diminishing the size of his kingdom. Despite service to the state, however, academicians had plenty of time for other studies, particularly several large communal projects they set for themselves, including exhaustive natural histories of plants and animals (Figure 17).

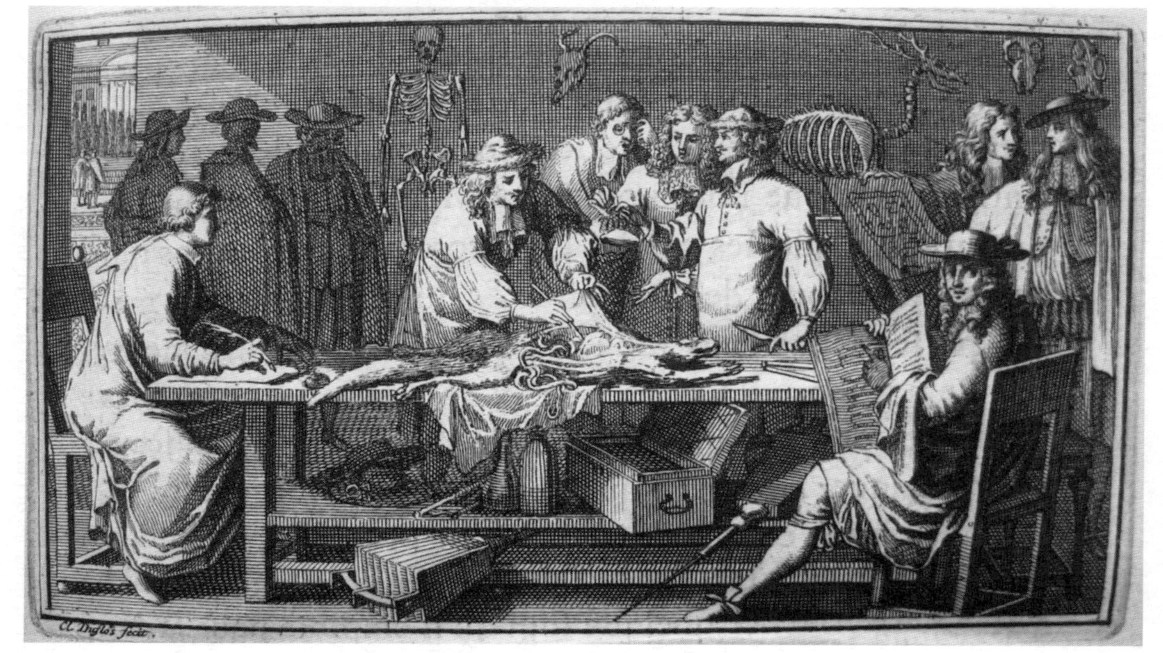

17. A dissection carried out by members of the Parisian Royal Academy of Sciences. The secretary (Jean-Baptiste Duhamel) records the observations while groups of academicians discuss them; the Jardin du Roi (King's Garden) is visible out of the window. *Mémoires pour servir à l'histoire des animaux*. (The Hague, 1731; originally published Paris, 1671)

Royal patronage also provided academicians with workspaces: a chemical laboratory, a botanical garden, and an astronomical observatory on the (then) outskirts of Paris. Completed in 1672, the Observatory of Paris was at first intended as a home for the entire Academy, but became the exclusive domain of the astronomers. The astronomer Gian Domenico Cassini, enticed away from the Pope's service to Paris by a huge stipend and control of the new observatory, took up residence there before the building was finished. He, and three generations of Cassinis after him, made the Observatoire the premier astronomical institution in Europe. Its north–south centreline marked the Earth's prime meridian from which longitude was widely measured for two centuries, until primacy was captured in 1884 by the line passing through Greenwich. (The Royal Observatory at Greenwich had been founded in 1675, shortly after the Paris Observatory, specifically for the 'finding out of the longitude of places for perfecting navigation and astronomy'.) Royal funding also allowed the Paris Academy to send scientific expeditions abroad – to Guyana, Nova Scotia, and Denmark for astronomical observations, to Greece and the Levant for collecting botanical specimens, and famously, in the early 18th century, to South America and Lapland to make observations and measurements to test Cartesian and Newtonian predictions for the exact shape of the Earth. It likewise collected and published observations sent by Jesuits from Siam, China, and elsewhere, and corresponded extensively with members of the Royal Society (even when France and England were at war) and other savants throughout Europe.

Scientific groups beyond the academies

Scientific academies proliferated after 1700, opening in Bologna, Uppsala, Berlin, St Petersburg, French provincial centres, and even at Philadelphia in the North American colonies, and became symbols of national pride and achievement. But academies were only one expression of the developing world of science. They were accompanied by more informal, but sometimes no less important, social groupings. In Paris, the Académie Royale followed upon natural philosophical *salons* held in private homes or public settings, where interested persons assembled for discussion, conversation, and debate under the leadership of an organizer. Their establishment testifies to how developments in natural philosophy had captured public attention, and was becoming a social phenomenon. In London, the new coffeehouses that opened in the later 17th century provided locales for a variety of people to meet and discuss issues, including natural philosophical ones. Public interest fuelled the emergence in the early 18th century of the public demonstrator, a character part natural philosopher and part showman who entertained and educated public gatherings (for an admission fee) with exotic apparatus or showy displays.

Less visible than the academies, but equally significant for the history of science were the networks of correspondence that linked individuals into webs of communication. Natural philosophers privately exchanged letters, manuscripts, and their newly printed books. The privacy of the letter allowed for the airing of unpopular and radically novel ideas, creating a mostly hidden discussion that carried on across Europe throughout the 17th century. This invisible 'republic of letters' (a Renaissance humanist phrase) united like-minded thinkers across national, linguistic, and confessional lines, and bridged the distances between them. The construction of such webs of correspondence was enhanced by people known as intelligencers. They received letters, organized and compiled the information, distributed it to interested parties, and sent out follow-up inquiries. The volume of a busy intelligencer's correspondence could be staggering. Nicolas-Claude Fabri de Peiresc (1580–1637), who encouraged Gassendi and spread Galileo's ideas in France, maintained about 500 correspondents and left behind over 10,000 letters. One of his correspondents, the Minim friar Marin Mersenne (1588–1648), was himself a communications hub. In his monastic cell in Paris, he received correspondence and disseminated the work of Descartes, Galileo, and others through a network across Europe.

In England, Samuel Hartlib (c. 1600–62), a Prussian refugee from the Thirty Years War, maintained correspondence linking all of Protestant Europe and North America; his 2,000 surviving letters are but a small fraction of what he wrote. Hartlib was motivated by Utopian and utilitarian ideas for the reform of education, agriculture, and industry after a Baconian fashion, but also by religious beliefs, particularly millenarian hopes for creating a Protestant 'paradise on earth' in England. His circle included entrepreneurs, moralists, natural philosophers, theologians, and engineers, and his projects ranged from opening technical colleges to improving brewing. The academies themselves became nodes in this epistolary network, and the learned journals – the *Philosophical Transactions*, the *Journal des Sçavans*, as well as their modern descendants – can be seen as formalized versions of it crystallized in print.

Thanks to the establishment of scientific academies and the increasing importance of technological applications in the 17th century, succeeding centuries saw a gradual professionalization of scientific work and a slow disappearance of the 'amateur' natural philosopher. Increased demand for knowledgeable, trustworthy people who could apply scientific knowledge and methods to practical problems drove the establishment of more formal and rigorous training for them in universities, and this in turn led to greater standardization of ideas and approaches. The cumulative result was the 19th-century emergence of 'science' as a career, of 'scientists' as a distinct social and vocational class (resembling in some respects what Bacon had described in the *New Atlantis*), and the gradual refashioning of the early modern world into the modern world of science and technology. That transformation was a slow and complex process, the account of which does not belong in this book. The turns upon the path that historical characters chose, the ideas and needs that influenced their decisions, and the events that enabled or disabled their intentions, were neither obvious nor preordained. While the realities of the natural world would be no different, the ways human beings express, conceptualize, and deploy them might very well be. The particular historical route we have chosen to tread has delivered us into a world of science and technology full of wonders to astonish the greatest advocates of *magia naturalis* and yet not without problems, both those remaining unsolved and those of our own making. Amid our enviable store of natural knowledge, the wise, peaceable, and orderly Bensalem continues to elude us, even if it has never ceased to inspire.

Epilogue

Virtually every text and artefact that has come down to us from early modern natural philosophers expresses their fervour to explore, invent, preserve, measure, collect, organize, and learn. Their innumerable theories, explanations, and world systems that jostled for recognition and acceptance met with various fates. Many early modern concepts and discoveries – Copernicus's heliocentrism, Harvey's circulation of the blood, Newton's inverse-square law of gravitation – constitute the foundations of our modern understanding of the world. Other ideas, like notions of atomism and estimates of the size of the universe, have been incrementally updated and refined by subsequent scientific work, and some, like Descartes' vortices or the mechanical explanation of magnetic attraction, have been discarded entirely.

Modern science continues to pursue many of the questions and aims of early modern natural philosophers – some of which they inherited from the Middle Ages, or even from the ancients. Like Gassendi, Descartes, and van Helmont, modern physicists continue to search for the ultimate particles of matter, to understand how these invisible bits of the universe unite and interact to form the world. Like Kepler, Cassini, and Riccioli, modern astronomers continue to scan and map the heavens, finding new objects and phenomena with instruments far more diverse and powerful than the quadrants and telescopes of Tycho, Galileo, or Hevelius. The explorers in New Spain like Hernández and Da Costa have heirs in scientists who continue to seek for new medicines in the plants and animals of jungles and deserts, or for new life forms in dark ocean trenches and even on distant worlds. Like their Paracelsian and chrysopoeian forebears, modern chemists labour to modify and improve natural substances and to create new materials, continuing the aspirations of Boyle to understand material change and of Bacon to provide things useful for human life. Like Vesalius, Malpighi, and Leeuwenhoek, modern biologists and physicians explore animal and human bodies with new instruments, uncovering ever finer structures and more astonishing mechanisms. Every new electronic gizmo that appears on the market refreshes the ties of technology to the wondrous and the magical.

Alongside such links of continuity, much has changed as well. The deep religious and devotional incentive that motivated early modern natural philosophers to study the Book of Nature – to find the Creator reflected in the creation – no longer provides a major driving force for scientific research. The constant awareness of history, of being part of a long and cumulative tradition of inquirers into nature, has been largely lost. Few scientists today would do as Kepler did when he subtitled his Copernican textbook 'a supplement to Aristotle', or seek for answers in ancient texts, where Newton sought for gravity's cause. The vision of a tightly interconnected cosmos has been fractured by the abandonment of questions of meaning and purpose, by narrowed perspectives and aims, and by a preference for a literalism ill-equipped to comprehend the analogy and metaphor fundamental to early modern thought. The natural philosopher and his broad scope of thought, activity, experience, and expertise has been supplanted by the professionalized, specialized, and technical scientist. The result is a scientific domain disconnected from the broader vistas of human culture and existence. It is impossible not to think ourselves the poorer for the loss of the comprehensive early modern vision, even while we are bound to acknowledge that modern scientific and technological development has enriched us with an astonishing level of material and intellectual wealth.

The Scientific Revolution was a period of both continuity and change, of innovation as well as tradition. The practitioners of early modern natural philosophy came from every part of Europe, every religious confession, every social background, and ranged from provocative innovators to cautious traditionalists. These disparate characters together contributed to the establishment of bodies of knowledge, institutions, and methodologies foundational to today's global world of science – a world that touches every living human being. We could tell them many things they were desperate to know, and they could perhaps in turn tell us things we are desperate to hear. Their age strikes us as both familiar and alien, simultaneously like our own and strikingly different. The very complexity and exuberance of the early modern period renders it the most fascinating and most important era in the entire history of science.

References

Chapter 1

Edward Grant, *The Foundations of Modern Science in the Middle Ages* (Cambridge: Cambridge University Press, 1996), p. 174.

Chapter 2

Giambattista della Porta, *Natural Magick* (London, 1658; reprint edn. New York: Basic Books, 1957), pp. 1–4.

Chapter 3

Nicholas Copernicus, *De revolutionibus* (Nuremberg, 1543), Schönberg's letter, fol. ii*r*; God's artisanship, fol. iii*v*; Osiander's 'preface', fols. i*v*–ii*r* (my translations). A full English translation is Copernicus, *On the Revolutions*, tr. Edward Rosen (Baltimore: Johns Hopkins University Press, 1992).

J. E. McGuire and P. M. Rattansi, 'Newton and the "Pipes of Pan"', *Notes and Records of the Royal Society*, 21 (1966): 108–43, on p. 126.

Chapter 4

Athanasius Kircher, *Mundus subterraneus* (Amsterdam, 1665), preface.

Galileo Galilei, *Il Saggiatore* [*The Assayer*], in *The Controversy on the Comets of 1618* (Philadelphia: University of Pennsylvania Press, 1960), pp. 183–4.

Chapter 6

The Works of Francis Bacon, ed. James Spedding, Robert L. Ellis, and Douglas D. Heath, 14 vols (London: 1857–74), 4:8, 3:294, 3:164.

Further reading

There are several good books surveying the Scientific Revolution in greater detail than is possible here. These include Peter Dear, *Revolutionizing the Sciences: European Knowledge and Its Ambitions, 1500–1700*, 2nd edn. (Princeton: Princeton University Press, 2009); John Henry, *The Scientific Revolution and the Origins of Modern Science*, 2nd edn. (Basingstoke: Palgrave, 2002); and Margaret J. Osler, *Reconfiguring the World: Nature, God, and Human Understanding from the Middle Ages to Early Modern Europe* (Baltimore: Johns Hopkins University Press, 2010). The last is especially good in providing technical details of early modern scientific ideas. A useful reference source is Wilbur Applebaum's *Encyclopedia of the Scientific Revolution* (New York: Garland, 2000), full of short, authoritative articles on hundreds of subjects.

Chapter 1

For the medieval (and ancient) background, see David C. Lindberg, *The Beginnings of Western Science*, 2nd edn. (Chicago: University of Chicago Press, 2007), and for a fascinating account of medieval voyages, see J. R. S. Phillips, *The Medieval Expansion of Europe*, 2nd edn. (Oxford: Clarendon Press, 1998). For Renaissance humanisms, see Anthony Grafton with April Shelford and Nancy Siraisi, *New Worlds, Ancient Texts: The Power of Tradition and the Shock of Discovery* (Cambridge, MA: Harvard University Press, 1992); and Jill Kraye (ed.), *Cambridge Companion to Renaissance Humanism* (Cambridge: Cambridge University Press, 1999). On other issues in this chapter, see Elizabeth Eisenstein,

The Printing Press as an Agent of Change (Cambridge: Cambridge University Press, 1979); Peter Marshall, *The Reformation: A Very Short Introduction* (Oxford: Oxford University Press, 2009); and Anthony Pagden, *European Encounters with the New World from the Renaissance to Romanticism* (New Haven: Yale University Press, 1993).

Chapter 2

On natural magic and its place in the history of science, see John Henry, 'The Fragmentation of Renaissance Occultism and the Decline of Magic', *History of Science*, 46 (2008): 1–48. On the background to the connected worldview, see Brian Copenhaver 'Natural Magic, Hermetism, and Occultism in Early Modern Science', pp. 261–301 in David C. Lindberg and Robert S. Westman (eds.), *Reappraisals of the Scientific Revolution* (Cambridge: Cambridge University Press, 1990). For an account of various sorts of *magia*, see D. P. Walker, *Spiritual and Demonic Magic: Ficino to Campanella* (University Park, PA: Pennsylvania State University Press, 1995). To correct widely held modern prejudices about the role of religion in science, see the very readable essays in Ronald Numbers (ed.), *Galileo Goes to Jail and Other Myths about Science and Religion* (Cambridge, MA: Harvard University Press, 2009), and for more in-depth treatments, David C. Lindberg and Ronald L. Numbers (eds.), *God and Nature: Historical Essays on the Encounter Between Christianity and Science* (Berkeley, CA: University of California Press, 1989).

Chapter 3

On the major characters discussed in this chapter, see Victor E. Thoren, *The Lord of Uraniborg: A Biography of Tycho Brahe* (Cambridge: Cambridge University Press, 1990); Maurice Finocchiaro (ed.), *The Essential Galileo* (Indianapolis, IN: Hackett, 2008); John Cottingham (ed.), *Cambridge Companion to Descartes* (Cambridge: Cambridge University Press, 1992); Richard S. Westfall, *The Life of Isaac Newton* (Cambridge: Cambridge University Press, 1994). For the best overview of the current understanding of 'Galileo and the Church', see the introduction to Finocchiaro, *The Galileo Affair* (Berkeley, CA: University of California Press, 1989). On astrology, see Anthony

Grafton, *Cardano's Cosmos: The World and Works of a Renaissance Astrologer* (Cambridge, MA: Harvard University Press, 1999). For better understanding of astronomical models and theories, see Michael J. Crowe, *Theories of the World: From Antiquity to the Copernican Revolution*, 2nd edn. (New York: Dover, 2001), and visit 'Ancient Planetary Model Animations' at http://people.sc.fsu.edu/~dduke/models.htm; created by Professor David Duke at Florida State University – this site contains outstanding animations of various planetary systems.

Chapter 4

For Galileo and motion, see the suggestions for Chapter 3.

For other major figures mentioned, see Alan Cutler (for Steno), *The Seashell on Mountaintop* (New York: Penguin, 2003); Paula Findlen (ed.), *Athanasius Kircher: The Last Man Who Knew Everything* (New York: Routledge, 2004); and Michael Hunter, *Robert Boyle: Between God and Science* (New Haven: Yale University Press, 2009). For alchemy and its importance, see Lawrence M. Principe, *The Secrets of Alchemy* (Chicago: Chicago University Press, 2011) and William R. Newman, *Atoms and Alchemy: Chymistry and the Experimental Origins of the Scientific Revolution* (Chicago: Chicago University Press, 2006). For a useful, but now rather dated, overview of the mechanical philosophy, see the relevant sections in Richard S. Westfall, *The Construction of Modern Science: Mechanisms and Mechanics* (Cambridge: Cambridge University Press, 1971).

Chapter 5

Nancy G. Siraisi, *Medieval and Early Renaissance Medicine* (Chicago: University of Chicago Press, 1990) and Roger French, *William Harvey's Natural Philosophy* (Cambridge: Cambridge University Press 1994). On natural history, see William B. Ashworth, 'Natural History and the Emblematic Worldview', in David C. Lindberg and Robert S. Westman (eds.), *Reappraisals of the Scientific Revolution* (Cambridge: Cambridge University Press, 1990), pp. 303–32; and Nicholas Jardine, James A. Secord, and Emma C. Spary (eds.), *The Cultures of Natural History* (Cambridge: Cambridge University Press, 1995). On the Spanish role, see María M. Portuondo, *Secret Science: Spanish Cosmography and the New World* (Chicago: University of Chicago Press, 2009) and Miguel de Asúa and Roger French, *A New World of Animals: Early Modern Europeans on the Creatures of Iberian America* (Burlington, VT: Ashgate, 2005).

Chapter 6

Pamela O. Long, *Technology, Society, and Culture in Late Medieval and Renaissance Europe, 1300–1600* (Washington, DC: American Historical Association, 2000); Paolo Rossi, *Philosophy, Technology, and the Arts in Early Modern Europe* (New York: Harper and Row, 1970); Markku Peltonen (ed.), *Cambridge Companion to Bacon* (Cambridge: Cambridge University Press, 1996); Lisa Jardine, *Ingenious Pursuits: Building the Scientific Revolution* (New York: Anchor Books, 2000); Marco Beretta, Antonio Clericuzio, and Lawrence M. Principe (eds.), *The Accademia del Cimento and its European Context* (Sagamore Beach, MA: Science History Publications, 2009); Alice Stroup, *A Company of Scientists: Botany, Patronage, and Community at the Seventeenth-Century Parisian Royal Academy of Sciences* (Berkeley, CA: University of California Press, 1990).

Index

A

B

C

D

E

F

G

M

N

O

P

R

S

T

GALILEO

A Very Short Introduction

Stillman Drake

Galileo's scientific method was of overwhelming significance for the development of modern physics, and led to a final parting of the ways between science and philosophy.

In a startling reinterpretation of the evidence, Stillman Drake advances the hypothesis that Galileo's trial and condemnation by the Inquisition in 1633 was caused not by his defiance of the Church, but by the hostility of contemporary philosophers.

Galileo's own beautifully lucid arguments are used to show how his scientific method was utterly divorced from the Aristotelian approach to physics in that it was based on a search not for causes but for laws.

'stimulating and very convincing'

Theology

www.oup.com/vsi

COSMOLOGY

A Very Short Introduction

Peter Coles

What happened in the Big Bang? How did galaxies form? Is the universe accelerating? What is 'dark matter'? What caused the ripples in the cosmic microwave background?

These are just some of the questions today's cosmologists are trying to answer. This book is an accesible and non-technical introduction to the history of cosmology and the latest developments in the field. It is the ideal starting point for anyone curious about the universe and how it began.

'A delightful and accessible introduction to modern cosmology'

Professor J. Silk, Oxford University

'a fast track through the history of our endlessly fascinating Universe, from then to now'

J. D. Barrow, Cambridge University

www.oup.com/vsi

65. World Music
66. Mathematics
67. Philosophy of Science
68. Cryptography
69. Quantum Theory
70. Spinoza
71. Choice Theory
72. Architecture
73. Poststructuralism
74. Postmodernism
75. Democracy
76. Empire
77. Fascism
78. Terrorism
79. Plato
80. Ethics
81. Emotion
82. Northern Ireland
83. Art Theory
84. Locke
85. Modern Ireland
86. Globalization
87. The Cold War
88. The History of Astronomy
89. Schizophrenia
90. The Earth
91. Engels
92. British Politics
93. Linguistics
94. The Celts
95. Ideology
96. Prehistory
97. Political Philosophy
98. Postcolonialism
99. Atheism
100. Evolution
101. Molecules
102. Art History
103. Presocratic Philosophy
104. The Elements
105. Dada and Surrealism
106. Egyptian Myth
107. Christian Art
108. Capitalism
109. Particle Physics
110. Free Will
111. Myth
112. Ancient Egypt
113. Hieroglyphs
114. Medical Ethics
115. Kafka
116. Anarchism
117. Ancient Warfare
118. Global Warming
119. Christianity
120. Modern Art
121. Consciousness
122. Foucault
123. The Spanish Civil War
124. The Marquis de Sade
125. Habermas
126. Socialism
127. Dreaming
128. Dinosaurs
129. Renaissance Art
130. Buddhist Ethics
131. Tragedy
132. Sikhism
133. The History of Time
134. Nationalism
135. The World Trade Organization
136. Design
137. The Vikings
138. Fossils
139. Journalism
140. The Crusades

210. Fashion
211. Forensic Science
212. Puritanism
213. The Reformation
214. Thomas Aquinas
215. Deserts
216. The Norman Conquest
217. Biblical Archaeology
218. The Reagan Revolution
219. The Book of Mormon
220. Islamic History
221. Privacy
222. Neoliberalism
223. Progressivism
224. Epidemiology
225. Information
226. The Laws of Thermodynamics
227. Innovation
228. Witchcraft
229. The New Testament
230. French Literature
231. Film Music
232. Druids
233. German Philosophy
234. Advertising
235. Forensic Psychology
236. Modernism
237. Leadership
238. Christian Ethics
239. Tocqueville
240. Landscapes and Geomorphology
241. Spanish Literature
242. Diplomacy
243. North American Indians
244. The U.S. Congress
245. Romanticism
246. Utopianism
247. The Blues
248. Keynes
249. English Literature
250. Agnosticism
251. Aristocracy
252. Martin Luther
253. Michael Faraday
254. Planets
255. Pentecostalism
256. Humanism
257. Folk Music
258. Late Antiquity
259. Genius
260. Numbers
261. Muhammad
262. Beauty
263. Critical Theory
264. Organizations
265. Early Music
266. The Scientific Revolution

141. Feminism
142. Human Evolution
143. The Dead Sea Scrolls
144. The Brain
145. Global Catastrophes
146. Contemporary Art
147. Philosophy of Law
148. The Renaissance
149. Anglicanism
150. The Roman Empire
151. Photography
152. Psychiatry
153. Existentialism
154. The First World War
155. Fundamentalism
156. Economics
157. International Migration
158. Newton
159. Chaos
160. African History
161. Racism
162. Kabbalah
163. Human Rights
164. International Relations
165. The American Presidency
166. The Great Depression and The New Deal
167. Classical Mythology
168. The New Testament as Literature
169. American Political Parties and Elections
170. Bestsellers
171. Geopolitics
172. Antisemitism
173. Game Theory
174. HIV/AIDS
175. Documentary Film
176. Modern China
177. The Quakers
178. German Literature
179. Nuclear Weapons
180. Law
181. The Old Testament
182. Galaxies
183. Mormonism
184. Religion in America
185. Geography
186. The Meaning of Life
187. Sexuality
188. Nelson Mandela
189. Science and Religion
190. Relativity
191. History of Medicine
192. Citizenship
193. The History of Life
194. Memory
195. Autism
196. Statistics
197. Scotland
198. Catholicism
199. The United Nations
200. Free Speech
201. The Apocryphal Gospels
202. Modern Japan
203. Lincoln
204. Superconductivity
205. Nothing
206. Biography
207. The Soviet Union
208. Writing and Script
209. Communism